T0177135

Origins of Life in the Universe

Origins of Life in the Universe traces the evolution of the Cosmos from the
Big Bang to the development of intelligent life on Earth. Conveying
clear, concise science in an engaging narrative it maps the history of the
Universe for introductory science and astrobiology courses for non-
science majors.

What is the origin of the Universe? How do stars and planets form? How
does life begin? How did intelligence arise? Are we alone in the Cosmos?
Physics, chemistry, biology, astronomy, and geology are combined to
answer some of the most fascinating questions in science and create a
chronicle of events in which the swirling vapors in the primordial cloud of
the Universe evolved over billions of years into conscious life. Coverage of
the latest discoveries in astrobiology give a sense of the excitement of this
fast-moving field.

ROBERT JASTROW, internationally acclaimed as an astronomer and
popularizer of science, was the founder of the Goddard Institute for Space
Studies, a US Government laboratory charged with carrying out research
in astronomy and planetary science. He has authored several books that
have made science accessible to a wider audience, and has been a frequent
commentator on science news. His research spanned nuclear physics,
planetary science, atmospheric physics, weather and climate prediction.

MICHAEL RAMPINO is an Associate Professor of earth science and
biology at New York University, and a Research Associate at GISS in New
York. His research spans many areas of the earth sciences, especially the
inter-relationships between Earth's changing environments and the
evolution of life. He is Series Co-Editor for the highly acclaimed
Encyclopedia of Earth Sciences (Springer) and the Editor-in-Chief of the *Earth
Science Encyclopedia On-Line.*

Origins of Life
in the Universe

ROBERT JASTROW   MICHAEL RAMPINO

CAMBRIDGE
UNIVERSITY PRESS

University Printing House, Cambridge CB2 8BS, United Kingdom

One Liberty Plaza, 20th Floor, New York, NY 10006, USA

477 Williamstown Road, Port Melbourne, VIC 3207, Australia

314–321, 3rd Floor, Plot 3, Splendor Forum, Jasola District Centre,
New Delhi – 110025, India

103 Penang Road, #05-06/07, Visioncrest Commercial, Singapore 238467

Cambridge University Press is part of the University of Cambridge.

It furthers the University's mission by disseminating knowledge in the pursuit of
education, learning and research at the highest international levels of excellence.

www.cambridge.org
Information on this title: www.cambridge.org/9780521825764

© R. Jastrow and M. Rampino 2008

First published 2008
Reprinted 2021

Printed in Great Britain by Ashford Colour Press Ltd.

A catalog record for this publication is available from the British Library

ISBN 978-0-521-82576-4 Hardback
ISBN 978-0-521-53283-9 Paperback

Contents

Extended contents

Foreword

For most of us, our daily lives are carried out within a space of a few miles – our lifetimes measured in a few tens of years. Science enters our lives as a supplier of ever-newer conveniences for living and of modern medical treatment if we become ill. A non-scientist might choose to take these things for granted, and get on with his or her life. Why then devote time and energy to the study of science, and in particular to astrobiology and the history of the Universe?

The answer is that astrobiology will introduce you to new visions of the Universe that are grand and inspiring. Beyond Earth, lies an endless parade of glittering stars and galaxies. Among them are worlds, some like our own, others strangely different, which may also harbor life, even intelligent life. A person who is unaware of these prospects would be similar to someone who attended a concert wearing earplugs, or who traveled to a land rich in beauty and history, but stayed entirely in their hotel room.

For those who wish to take their minds on a more adventurous journey, this text will provide an admirable guide. I have known Bob Jastrow and Mike Rampino for years and admired the breadth of their interests, which range from the birth of stars to the extinction of the dinosaurs. I commend you into their expert care, and trust that you will have a magnificent trip.

Robert Shapiro
Professor Emeritus of Chemistry, New York University

Preface

Origins of Life in the Universe deals with the developing field of astrobiology – the study of the scientific disciplines bearing on the emergence of life and intelligence in the Cosmos. It combines material from the traditional disciplines of physics, chemistry, biology, astronomy, and geology to create a chronicle of events in which swirling vapors in the primordial Universe evolved over billions of years into conscious life on Earth.

The book is written to support an introductory science course aimed specifically at non-science students. In it, the basic facts and concepts of the major scientific disciplines are taught as a connected narrative – the story of the evolving Universe from the Big Bang to the appearance of intelligent life. In this book, facts and concepts in separate disciplines are taken up in a natural sequence as needed to advance the narrative. The book addresses some of the most fascinating and fundamental questions in science: What is the origin of the Universe? How do stars and planets form? How does life begin? How did intelligence arise? Are we alone in the Cosmos?

According to the scientific evidence, the story began approximately 14 billion years ago, when the Universe expanded in the aftermath of the cosmic explosion known to astronomers as the Big Bang. In a material sense, the seed of everything that has happened in the last 14 billion years can be traced back to the moment of the Big Bang; every star, planet, and living creature in the Universe came into being as a result of events set in motion at that moment.

What was the cause of the cosmic explosion? What was the Universe like before the explosion occurred? Did the Universe even exist before that moment? These questions are at the forefront of astrophysical research. The answers remain among the greatest of scientific mysteries. In contrast, the collective labors of scientists in the modern era have yielded a detailed account of the events that followed the cosmic explosion. The Universe expanded rapidly outward from a hot, dense state, cooling as it expanded. Initially there were no galaxies, stars, or planets. But after about a billion years, the expanding cloud of matter and energy had cooled sufficiently to permit the formation of condensed knots of matter held together by gravity. These were the first galaxies and stars.

Another 8 or 9 billion years went by, and then, some 4.5 billion years ago, the Sun and its planets, including Earth, appeared as inconspicuous condensations in an undistinguished spiral galaxy – one among billions of galaxies, each containing hundreds of billions of stars.

At some point in the first billion years after Earth formed, life appeared in the waters covering its surface. We know this from the fossil record. How was that life created out of inanimate matter? A number of plausible theories exist, but we still cannot say which, if any, of these ideas is correct.

Once across the threshold of life, the scientist picks up the trail again. The remains of simple, bacteria-like organisms appear first in the record of the rocks about 3.5 billion years ago; traces of soft-bodied many-celled organisms are preserved in rocks about 600 million years old. A little later, about 540 million years ago, the remains of the first marine animals with hard shells and skeletons appear in the record. From this point on, the record of the history of life is fairly complete, until finally the remains of the first creature with a human level of intelligence are found in rocks about one hundred and fifty thousand years old.

All the larger trends in this history are clearly exhibited in the fossil record. This is particularly true of human ancestors, of whom essentially nothing was known when Charles Darwin first developed his theory of evolution by natural selection. Discoveries in recent years have filled in much of the missing record of human origins. Some gaps exist, but new finds are made nearly every year.

As scientists, we have both been involved in research in fields related to astrobiology. Jastrow, as director of NASA's

Goddard Institute for Space Studies and The Mount Wilson Observatory, directed cutting-edge research in astronomy, planetary science, and geology, and was the first chairman of NASA's Lunar Exploration Committee. Rampino has done geological fieldwork on six continents investigating volcanic activity, the record of climatic change, and role of catastrophic events in the history of life.

We hope that this book and the story it tells will convey to students the excitement and sense of wonder that we feel when we contemplate our long cosmic history. Our combined experience, based on teaching introductory college science courses over a period of 25 years, indicates that this chronological narrative of the evolution of the Cosmos strongly appeals to students with a wide variety of backgrounds and interests.

During this time, we have also developed supporting materials, including end-of-chapter questions with answers, and additional questions at www.cambridge.org/jastrow.

Many colleagues helped us with the preparation of this volume. We are particularly indebted to several people who were kind enough to take the time required for a careful reading and the preparation of detailed criticisms of individual chapters, and eventually the entire text. These include Dr. Richard Stothers and Dr. Andre Adler for the chapters on the contents of the Universe, cosmology, and the birth and death of stars; Dr. Michael Gaffey for the chapter on the Solar System; Dr. Christian Koeberl for the chapter on the Moon; Dr. Vivien Gornitz for the chapters on Mars and Venus; Dr. Dennis Kent for the chapters on the geology of Earth and plate tectonics; Drs. Andre Lapenis and Tyler Volk for the chapter on climate; Drs. Robert Shapiro, Edward Berger, and Stephen Small for the chapter on the origin of life and evolution (Dr. Shapiro made considerable contributions to that chapter); Dr. Nick Butterfield for the chapter on the early history of life; Dr. David Varricchio for the chapters on the dinosaurs and mammals; Dr. Donald Johansson for the chapter on human evolution; and Dr. Robert Shapiro for the chapter on life in the Cosmos.

Earlier versions of the manuscript were used as class notes for introductory level science courses at Columbia University,

Dartmouth College, and New York University. We thank our students and teaching assistants for their many comments and criticisms of the material as we continued to refine it.

We are especially indebted to Matt Lloyd, our editor at Cambridge University Press, whose detailed reading and criticism of the entire manuscript strengthened the book enormously.

PART I

The Universe

CHAPTER 1

Our place in the Universe

The realm of the galaxies

All life as we know it exists within the bounds of the single planet that we call home. For centuries mankind has gazed towards the heavens and wondered whether there may be others like ourselves, living on distant worlds like our own. Only recently have technological advances allowed us to begin gathering information that may provide some clues. The vastness of space is difficult for the human mind to comprehend, as such enormous sizes and distances are beyond those we encounter in our daily lives. Yet understanding how Earth fits into the grand scale of the Universe provides us with an important new perspective on life.

To us Earth is a vast planet, yet on the scale of the Universe we are a mere grain of sand. This chapter will provide an important sense of scale by explaining the size and contents of the Universe. Beginning at our own Solar System, we will take a look at the Sun and the relative sizes of the planets in orbit around it, using analogies with everyday objects and distances that we can understand more easily. We then move outwards to our nearest stars, in reality millions of miles away yet still close in terms of the Universe as a whole. Gravity holds together the group of stars closest to us, forming our Galaxy, the Milky Way. Contemplating increasingly larger distances, we will see that the Milky Way is only one galaxy in a cluster of galaxies, and that this cluster is only one of an infinite number of clusters in our Universe.

1.1 The Sun and planets

The Sun, its family of planets with their moons, and a large number of smaller bodies, form the *Solar System*. Earth, the third planet from the Sun, travels around the Sun at an average distance of 93 million miles (150 million kilometers).[1] This distance is defined as one astronomical unit or AU. Between Earth and the Sun lie the planets Mercury and Venus. Outside our orbit lie the planet Mars, the giant planets Jupiter, Saturn, Uranus, and Neptune, and the frozen world of Pluto (see Figure 1.1). Pluto, the outermost planet, travels in an elliptical orbit at distances from the Sun ranging between 3 and 4 billion miles (5 and 6 billion kilometers), some 30 to 40 AU.

Pluto now appears to be the largest of a large group of objects, called the Kuiper Belt, which lies in a zone surrounding the Sun. Far beyond the orbit of Pluto are the comets, small icy bodies that surround the Sun in a spherical halo called the Oort Cloud. Beyond that, space contains nothing but a few atoms of hydrogen per cubic centimeter until we reach the stars that are the Sun's neighbors. These

---

[1] In this book, we use the Imperial System units first, with the metric system units in parenthesis.

Previous page: Hubble Ultra Deep Field infrared view of galaxies billions of light-years away.
  NASA, ESA, and R. Thompson (University of Arizona)

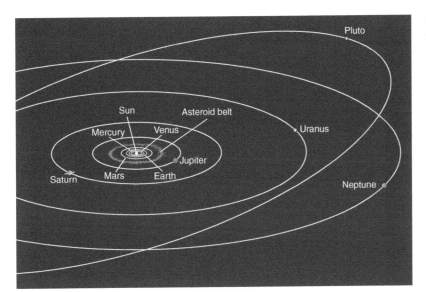

FIGURE 1.1 The Solar System.

nearest stars are approximately 30 trillion miles (50 trillion kilometers) away.

An analogy will help to clarify the meaning of such enormous distances. Let the Sun be the size of an orange; on that scale Earth is a grain of sand circling in orbit around the Sun at a distance of 30 feet (10 meters); Jupiter, 11 times larger than Earth, is a cherry pit revolving at a distance of 200 feet (65 meters) or one city block; Saturn is another cherry pit two blocks from the Sun; and Pluto is still another smaller grain of sand at a distance of ten city blocks from the Sun. The nearest stars are orange-sized objects more than a thousand miles (1500 kilometers) away.

An orange, a few grains of sand close by, and some cherry pits circling slowly around the orange at a distance of a city block; more than a thousand miles (1500 kilometers) away is another orange, perhaps with a few specks of planetary matter circling around it. That is the void of space.

1.2 The Sun's nearest neighbors

The Sun's closest neighbor is the star Alpha Centauri. Alpha Centauri is 24 trillion miles (38 trillion kilometers) from our Solar System, or slightly closer than the average distance between stars in our neighborhood. It is actually a triple star – a family of three stars formed simultaneously out of a single cloud of gas and dust. Ever since their birth, the three stars have circled one another under the attraction of gravity.

The largest of the three stars in Alpha Centauri resembles the Sun and possesses a similar surface temperature and color. The middle-sized star of the triplet, somewhat smaller than the Sun and orange in color, circles around the largest star in a close waltz at a distance of 2 billion miles (3.2 billion kilometers). One turn around takes 80 years for this pair. The third member is a very small, faint, red star, a tenth as large as the Sun, which circles the two other members of the triplet at a distance of a trillion miles (1.6 trillion kilometers), completing one turn in a million years.

The Alpha Centauri group are the closest stars to us that are bright enough to be visible to the naked eye. However, the Sun may have still closer neighbors – very small, dimly luminous stars – too faint to have been detected thus far. There may also be burned-out stars that have exhausted their fuel in the space between the Sun and Alpha Centauri. Finally, there may be many bodies the size of planets, too small to glow by their own energy, in the space around us. All these possibilities await the future exploration of the regions outside the Solar System.

1.3 Our Galaxy

The Sun and its neighbors are only a few among approximately 400 billion stars that are bound together by gravity in an enormous cluster, called the *Galaxy*. Most, if not all, stars in the Universe are held within such clusters. These other clusters also are called galaxies.

The stars in the Galaxy revolve about its center as the planets revolve about the Sun. The Sun itself participates in this motion, completing one circuit around the Galaxy in about 225 million years. The Galaxy is flattened by its rotating motion into the shape of a disk, whose thickness is roughly one-twentieth of its diameter. The Sun is located within the disk about three-fifths of the way out from the center to the

FIGURE 1.2 The Milky Way Galaxy viewed edge-on. This image of the Galaxy was obtained with the Cosmic Background Explorer (COBE) satellite in the infrared part of the spectrum. The thin disk of the Galaxy and the central bulge are visible because obscuring dust is transparent in that region of the spectrum. The arrow indicates the position of the Sun in our Galaxy.

NASA Goddard Space Flight Center

edge. A spherical clump of closely spaced stars, called the central bulge of the Galaxy, surrounds the galactic nucleus at the center of the disk (Figure 1.2).

The general appearance of the Galaxy is shown clearly in photographs of other galaxies that are similar to ours and that happen to be oriented in space so that we see them at different angles. If you could stand outside the Galaxy and view it edge-on, it would look very much like NGC 4013[2] (Figure 1.3), with a layer of dust and gas clearly visible in the plane of the Galaxy. Viewed face-on, the Galaxy would look like NGC 3310 (Figure 1.4), with spiral arms in which most of the bright stars of the Galaxy (including the Sun) are located.

Our relative position within the Galaxy is evident when we look up at the night sky and see a luminous band stretching across the sky – the Milky Way (Figure 1.5). This band is composed of so many stars that they are not visible as separate points of light but blend together. The irregular dark lanes running through the center of the Milky Way are caused by

---

[2] NGC: New General Catalogue, a catalog of extended objects – galaxies, star clusters, and nebulas – compiled by Cambridge University astronomers in 1890. Another such catalog is the Messier Catalogue (M), an earlier compilation published by Charles Messier, a French astronomer, in 1784.

FIGURE 1.3 Edge-on view of the spiral galaxy NGC 4013 that lies about 55 million light-years from Earth. The image, taken by the Hubble Space Telescope, shows huge clouds of dust and gas extending along the galaxy's disk.

NASA Jet Propulsion Laboratory (NASA-JPL)

the extensive clouds of fine dust, concentrated in the central plane of the Galaxy, which block out the light of many stars.

An inspection of the Milky Way with even a modest-size telescope reveals the immensity of the number of stars concentrated in the Galaxy. In Figure 1.6, tens of thousands of stars can be seen, although only a very small portion of the night sky is shown.

The vastness of the space between the stars is difficult to comprehend. The same analogy that we used to clarify the meaning of the size of the Solar System is helpful in attempting to comprehend the emptiness of the Galaxy. Suppose again that the Sun is reduced from its million-mile (1.6 million-kilometer) diameter to the size of an orange. The Galaxy, on this scale, is a cluster of 400 billion oranges, each orange separated from its neighbors by an average distance of more than 1000 miles (1600 kilometers). In the space between, there is nothing but a tenuous distribution of atoms and a few

FIGURE 1.4 The spiral galaxy NGC 3310, resembling our Galaxy viewed face on.

NASA and The Hubble Heritage Team (STScI/AURA)

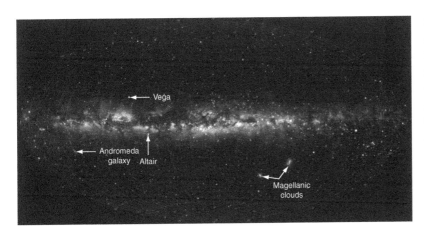

FIGURE 1.5 The Milky Way: An edge-on view of our Galaxy as seen from the inside. In this image, the whole of the sky is projected onto a two-dimensional map. The Magellanic Clouds, visible only in the Southern Hemisphere, are two dwarf galaxies anchored to our Galaxy by its gravitational force, each containing only a few billion stars.

Lund Observatory, Sweden

FIGURE 1.6 A small region of the night sky showing a background of tens of thousands of stars in our own Galaxy.

© Anglo-Australian Observatory/David Malin Images

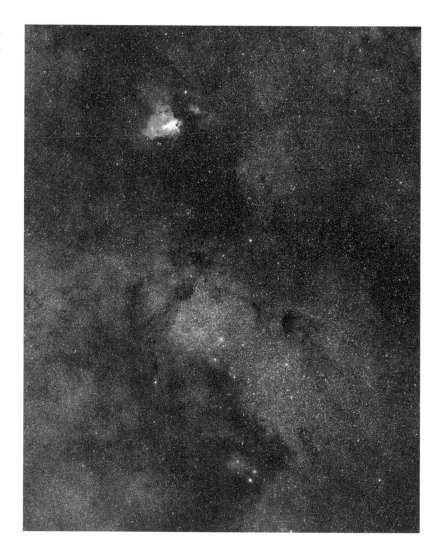

molecules and dust grains. That is the emptiness of space in the Galaxy.

The stars within the Galaxy are separated from one another by an average distance of 30 trillion miles (50 trillion kilometers). To avoid the frequent repetition of such awkwardly large numbers, astronomical distances are usually expressed in units of the light-year, defined as the distance covered in one year by a ray of light traveling 186 000 miles (300 000 kilometers) per second. This distance turns out to be approximately 5.9 trillion miles (9.5 trillion kilometers); hence, in these units, the distance from the Sun to Alpha Centauri is 4.3 light-years, the average distance between the stars in the

Galaxy is 5 light-years, and the diameter of the Galaxy is about 100 000 light-years.

1.4 Neighboring galaxies

Although the stars within our Galaxy are very thinly scattered, they are, nonetheless, relatively close together in comparison to the space that separates our Galaxy from neighboring galaxies. The distance to the next nearest galaxy comparable in size to ours is more than 2 million light-years, or 20 times the diameter of our Galaxy. It is difficult to imagine the emptiness of intergalactic space. Once outside the Galaxy, we encounter a region empty of stars and nearly empty of gas and dust.

No vacuum ever achieved on Earth can match the vacuum of the space outside our Galaxy. But if we go far enough away from the Galaxy, we come to other galaxies, collections of billions of stars held together, like ours, by the force of gravity. These galaxies are island universes – isolated clusters containing vast numbers of stars – each separated from the others by the void of intergalactic space.

The closest large galaxy, similar to the Milky Way Galaxy in size, is the Andromeda Galaxy, which is about 2.2 million light-years away from us. This galaxy happens to resemble our own closely in size and shape; it is a disk-shaped spiral of stars, gas, and dust, containing a few hundred billion stars in all, the entire collection of matter slowly spinning around a central axis like a gigantic pinwheel (Figure 1.7).

Andromeda is the only major galaxy visible to the naked eye, and it is the most distant object that can be seen without the aid of a telescope. However, it is not conspicuous, despite the fact that its intrinsic brilliance is roughly 100 billion times that of the Sun. Because of its enormous distance, Andromeda is barely visible to the naked eye, under the best conditions, as a very faint patch of light. But if it is photographed with even a modest-sized telescope, the faint patch is seen to have a structure that resembles our Galaxy, with a brightly glowing center, a distinct impression of spiral arms, and dark lanes formed by obscuring clouds of dust.

FIGURE 1.7 The Andromeda Galaxy (M31), our nearest galactic neighbor comparable in size to the Milky Way Galaxy. The arrows point to two satellite galaxies close to Andromeda. Jason Ware

1.5 Clusters of galaxies

Galaxies are grouped together into clusters of galaxies. The Milky Way Galaxy and Andromeda Galaxy, for example, are members of a small cluster of about 30 galaxies called the Local Group. In 1994, a new member of the Local Group, a small elliptical galaxy named the Sagittarius Dwarf Galaxy, was discovered only about 80 000 light-years from the Milky Way. This small, faint galaxy may be the nearest small galaxy to our own.

About 500 million light-years from our Galaxy, in the direction marked by the constellation Hercules, is a giant group of galaxies called the Hercules Cluster (Figure 1.8), which contains about 10 000 galaxies, each with 10 billion to 100 billion stars. Clusters of galaxies are grouped together in huge associations called *superclusters*. These superclusters can contain tens of clusters spread across more than 100 million light-years. At the largest scale, the clusters and superclusters of galaxies are not uniformly distributed in space, but are stretched out into threadlike filaments separated by enormous voids (Figure 1.9).

FIGURE 1.8 A small region of the Hercules cluster of galaxies, roughly 500 million light-years from Earth, containing about 50 galaxies. Nearly every spot of light in this photograph is a galaxy containing billions of stars.
NOAO/AURA/NSF

1.6 The observable Universe

The largest telescopes can photograph galaxies that are billions of light-years from Earth. If we built still larger telescopes, could we see farther out to space? Is there any limit to the range of our observations? A surprising answer has come out of recent astronomical discoveries.

According to the latest evidence, the Universe is approximately 14 billion years old. Therefore, we can only see galaxies that are less than 14 billion light-years away from us; if a galaxy is more than 14 billion light-years away, the light from it has not yet had time to reach Earth. Fourteen billion

FIGURE 1.9 The latest map of the Universe (prepared by the Sloan Digital Sky Survey). The galaxies are first identified in two-dimensional images (right) and their distances from Earth are estimated. This produces a three-dimensional map (left) showing the distribution of galaxies in the Universe out to 2 billion light-years.
  Jeffrey Newman, Lawrence Berkeley National Laboratory

light-years is the radius of the *observable Universe*. The true Universe may be infinitely large and contain an infinite number of galaxies and stars, but we can never hope to see them all from Earth.

1.7 The composition of the Universe

All matter in the Universe is made of the same basic building blocks. Ninety-two naturally occurring elements are known (Appendix B). The elements consist of atoms, each with a nucleus surrounded by orbiting electrons. The nucleus is a compact ball of protons and neutrons held together by the nuclear force. Each element from one to ninety-two is defined by the number of positively charged protons in its nucleus. Negatively charged electrons orbiting the nucleus are held in orbit by the electromagnetic force, which is the force of electrical attraction between two particles with unlike charges.

The chemical properties of the elements are determined by the number and arrangement of the electrons in their atoms. The number of electrons in an atom is equal to the number of protons in its nucleus. Thus, the number of protons in the nucleus of an element is its quintessential property. The neutrons in the nucleus help hold the nucleus together and serve to make the nucleus and the atom heavier. Two elements with the same number of protons in the nucleus (therefore chemically identical), but with a different number of neutrons, are called *isotopes*.

The simplest atom has a nucleus consisting of one proton, surrounded by one orbiting electron. This is the element hydrogen. It is the simplest and also the most abundant element in the Universe. Next is an atom whose nucleus consists of one proton and one neutron and orbited by one electron. This is heavy hydrogen, an isotope of hydrogen. The nucleus of heavy hydrogen is called the deuteron; the atom formed from it by adding an electron to the deuteron is called deuterium. There is also a nucleus made of one proton and two neutrons, called the triton; the atom formed from it is called tritium and is another isotope of hydrogen.

The next element beyond hydrogen has four particles in its nucleus, two protons and two neutrons, plus two orbiting electrons. It is called helium. There is also a light isotope of helium, whose nucleus has two protons and only one neutron.

Continuing in this way, we would expect to find a nucleus with five particles in it: three protons and two neutrons, or two protons and three neutrons. There is no such nucleus, however, because five particles do not hold together to form a stable nucleus. The next nucleus that actually exists has six particles in its nucleus, three protons and three neutrons. This is a nucleus of the element lithium.

Carbon is the name of the element with a nucleus consisting of six protons and six neutrons, with six electrons orbiting. Carbon also has isotopes with seven and eight neutrons in the nucleus. The element oxygen has a nucleus with eight protons. The two most common isotopes of oxygen have eight and ten neutrons respectively. Uranium, the heaviest naturally occurring element, has 92 protons in the nucleus and occurs as isotopes with 143 and 146 neutrons.

1.8 The basic forces of nature

All objects in the Universe, from the smallest atomic nucleus to the largest galaxy, are held together by only three fundamental forces – the nuclear force, the force of electromagnetism, and the force of gravity.

Most powerful is the *nuclear force* – the strong force of attraction that binds neutrons and protons together into atomic nuclei. Until recently, it was believed that a fourth independent force existed, called the *weak force*, which plays a role in nuclear reactions such as radioactive decay, involving electrons, positrons, and neutrinos. However, recent evidence indicates that the weak force and the electromagnetic force are two manifestations of a single basic force (called the *electroweak force*). The nuclear force of attraction pulls the particles of the nucleus together into a very compact body with a density of 1 billion tons per cubic inch. Although it is an exceedingly strong force, it has a very short range. The nuclear force will not attract two particles strongly if they are more than one ten-trillionth of an inch apart.

The *electromagnetic force* is more complicated, because its action on some pairs of particles is to pull them together, while on others its action is to push them apart. The explanation of these two kinds of action has come out of laboratory studies of electricity during the last 200 years. These studies show that two types of electric charge exist, positive and negative. Protons, for example, carry a positive charge, and electrons carry a negative charge. The studies also show that two particles bearing the same kind of charge – that is, both positive or both negative – repel each other. If two particles carry opposite charges, they attract one another. The nucleus of the atom, which carries positively charged protons, can hold electrons in orbit around it for this reason. The strength of the electromagnetic force falls off with the square of the distance between the two charged objects.

Least powerful is the force of *gravity*. Gravity acts between particles that have mass, and is always an attractive force, although it falls off rapidly with the square of the distance. Because it acts over all distances and is always attractive, gravity holds the Moon in orbit around Earth, Earth and other planets in revolution around the Sun, the Sun and other stars clustered together in our Galaxy, and galaxies grouped together in clusters and superclusters throughout the Universe.

1.9 Summary

By the end of this chapter you should be familiar with the following concepts and topics:

Size and contents of the Universe
The Solar System
  The Sun
  Relative sizes of the planets in orbit around the Sun
Our nearest stars
Our Galaxy, the Milky Way
One galaxy in a cluster of galaxies
Infinite number of clusters in our Universe
The composition of the Universe
The basic forces of nature

Questions

1.1. What is a galaxy? Draw the shape of a typical spiral galaxy. How large is a typical galaxy (e.g., the Milky Way) in light-years?

1.2. Compute the length of a light-year in miles.

1.3. What is the average distance in miles and light-years between stars in our Galaxy? What is the distance in miles and light-years between our Galaxy and the Andromeda Galaxy?

1.4. What are the basic building blocks of matter? What basic forces act between them? What are the main properties of these forces?

CHAPTER 2

A view of the origin of the Universe

Evidence for an explosive beginning

Before we can understand how life arose on Earth, we must first look back to the beginnings of time, to understand how the Universe began. Modern humans have only been around for a tiny fraction of the time that the Universe has been in existence, and understanding events in terms of a cosmic timescale is a great challenge for human minds. Yet our unique ability to contemplate our own origins has enabled us to trace our existence back to one specific event that occurred billions of years ago – the Big Bang. The conditions had to be exactly right from the very beginning, enabling the components of our Galaxy to form, ultimately resulting in life. Fourteen billion years ago the evolution of the Universe had begun, and the vast timescales on which things were formed was set in motion.

For almost a century, scientists have puzzled over observations and tried to understand the events that have led to the current state of our Universe. Outlining major discoveries by prominent astronomers, this chapter uncovers current theories relating to the origin and evolution of the Universe. The theory of the origin of the Universe is widely accepted and we will study this through analogies with everyday events to explain the expansion of the Universe. From this rate of expansion, the "age" of the Universe can also be determined. We will see how evidence supporting the Big Bang theory has raised some interesting questions, which ultimately impact directly on our own existence. Why should the density of matter have just the right value favorable for the emergence of a universe containing living beings? Since we exist within a universe that is still evolving, what is the fate of the Universe and everything within it?

2.1 Evidence for the explosive origin of the Universe

Thousands of galaxies exist within a distance of 100 million light-years from us. Around 1913, the American astronomer Vesto Melvin Slipher made a study of the speeds with which galaxies move through space. He found that most of the galaxies within range of his telescope were traveling at very high speeds, in some cases as much as several million miles (kilometers) an hour. Furthermore, *nearly all these galaxies were moving away from Earth.*

The fact that galaxies rushed across the heavens at tremendous speeds was a surprise to Slipher. The discovery that all except a few nearby galaxies were moving away from Earth was an even greater surprise. Our planet and its parent star, the Sun, are humble bodies in the cosmic hierarchy. Why should all the great star-clusters in the sky retreat from our neighborhood? It would be more reasonable to expect them to move in random directions; then, according to the laws of chance, at any particular moment roughly half the galaxies in the Universe would be moving toward Earth and half would be moving away from it. But Slipher's measurements indicated that this was not the

Previous page: Distant spiral galaxy NGC 4603, home to variable stars.
Jeffrey Newman (University of California at Berkeley) and NASA

case. If his measurements were correct, the entire Universe was moving away from one special point in space, and Earth was located at that point.

After some years of research on the galaxies, Slipher dropped this line of work and turned to other astronomical problems. At that time, another great astronomer, Edwin Hubble, entered the picture. Hubble turned the power of the 100-inch (2.5-metre) telescope on Mount Wilson in California, then the world's largest, on the problem of the moving galaxies.

Hubble used the 100-inch telescope to measure the speeds of many other spiral galaxies that were too faint to have been seen by Slipher with his modest-sized instrument. He confirmed Slipher's discovery and found that all the distant galaxies in the heavens were moving away from us at high speeds. The most distant galaxy he could observe was retreating from Earth at the extraordinary velocity of 100 million miles (150 million kilometers) an hour.

After the Second World War, the great power of the 200-inch (5-meter) telescope on Mount Palomar, also in California (Figure 2.1) was brought to bear on the problem of the receding galaxies and again Slipher's discovery was confirmed; every galaxy within the range of this mammoth instrument was retreating from Earth at an enormous speed.

Many more measurements have been made, down to the present day, and no important exception has been found to the rule discovered by Slipher. Regardless of the direction in which we look out into space, the distant galaxies in the Universe are moving away from us and from one another. The Universe is blowing up before our eyes, as if we are witnessing the aftermath of a gigantic explosion. This picture of the origin of the Universe, which rests on the discovery of the outward motions of the galaxies, is called the "Big Bang" theory.

2.1.1 How the speeds of galaxies are measured: The Doppler shift

How did Slipher and Hubble measure the speeds of distant galaxies? It is impossible to make such measurements directly by tracking a galaxy across the sky because the great distances to these objects render their motions imperceptible when

FIGURE 2.1 Edwin Hubble observing at the prime focus of the 200-inch (5-meter) telescope at Mount Palomar, California.

Palomar observatory, submillimeter observatory and California Institute of Technology

they are observed from night to night, or even from year to year. The closest spiral galaxy to us, Andromeda, would have to be observed for 500 years before it moved a measurable distance across the sky.

The method used by astronomers is indirect, and depends on the fact that when a galaxy moves away from Earth, its color changes. The effect occurs because light is a train of waves in space. When the source of the light moves away from the observer, the waves are stretched or lengthened by the receding motion. The length of a light wave is perceived by the eye as its color; short waves create the sensation that we call "blue," while long waves create the sensation of "red." Thus, the increase in

This observer sees blueshift

This observer sees redshift

FIGURE 2.2 The Doppler effect. Wavelength is affected by motion between the light source and an observer. Light waves are being emitted from a source of light that is moving toward the left (direction of the arrow). Note that the waves are crowded together in front of the moving source but are spread out behind it. Consequently, wavelengths appear shortened (blue-shifted) if the source is moving toward the observer. They appear lengthened (red-shifted) if the source is moving away from the observer.

the length of the light waves coming from a receding object is perceived as a reddening effect. The degree of the color change is proportional to the speed of the galaxy. This effect is called the *Doppler shift* (Figure 2.2).

If the galaxy is moving away from Earth, the Doppler shift makes its color seem redder than normal. This effect is called a *red shift*. If the galaxy is moving toward Earth, its color becomes bluer than normal. This is called a *blue shift*. All the distant galaxies studied by Slipher, and later by Hubble, showed a reddening, or red shift. This red shift, which betrayed the retreating movements of the galaxies, was the basis for the picture of the expanding Universe.

How is the red shift itself measured? First, a device similar to a prism is attached to a telescope. The prism spreads out the light from the moving galaxy into a band of colors like a rainbow. This band of colors is called a spectrum. In the next step, the spectrum is recorded on a photographic plate. Finally, the spectrum of the galaxy is lined up alongside the spectrum of a non-moving source of light. The comparison of the two spectra determines the red shift.

Figure 2.3 shows how the method works. The photographic images of several galaxies are shown at left, while the spectra of the same galaxies, recorded photographically, appear at right as tapering bands of light.

The spectra of the galaxies are rather indistinct because the galaxies are faint and far away. However, each spectrum contains one important feature. This is the pair of dark lines indicated by the arrow in the top spectrum. The lines represent colors created by glowing atoms of calcium in stars within the galaxy, which make useful markers for determining the amount of the red shift in a galaxy's spectrum.

The arrow points to the position the calcium colors normally would have in the galaxy's spectrum, if this galaxy were not

Galaxy, part of cluster in:	Distance (light-years)	H + K Red shifts	

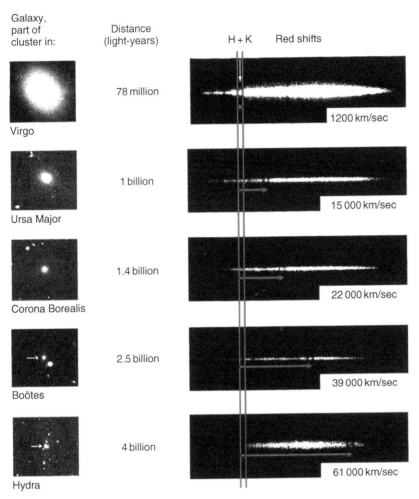

Virgo — 78 million — 1200 km/sec

Ursa Major — 1 billion — 15 000 km/sec

Corona Borealis — 1.4 billion — 22 000 km/sec

Boötes — 2.5 billion — 39 000 km/sec

Hydra — 4 billion — 61 000 km/sec

FIGURE 2.3 Galaxies at various distances and their spectra, showing the corresponding red shifts as indicated by the length of the red arrow (1 megaparsec = 3.26 million light-years).

Palomar Observatory, submillimeter observatory and California Institute of Technology

moving away from us. The length of the red arrow is the amount of the red shift.

The topmost photograph shows a galaxy that is about 78 million light-years from us. It is close enough to appear as a large, luminous shape, but too distant for us to see its individual stars. The calcium colors in its spectrum are shifted toward the red by a small but significant amount. The speed of this retreating galaxy, calculated from its red shift, turns out to be about 3 million miles an hour (1200 kilometers per second).

The next galaxy is a billion light-years away, and correspondingly smaller and fainter. The position of the calcium colors in its spectrum reveals a much greater shift toward the red, indicating a greater velocity of recession. The red shift in the spectrum of this galaxy corresponds to a speed of 126 million miles an hour (15 000 kilometers per second). The next three galaxies are 1.4 billion, 2.5 billion and 4 billion light-years away, and their velocities are 50 million, 90 million and 136 million miles per hour (22 000, 39 000 and 61 000 kilometers per second), respectively. More distant galaxies have been found to be moving away from us even more rapidly.

2.2 The law of the expanding Universe

In the 1920s, when Hubble began work on the spiral galaxies, astronomers did not know what these luminous spirals were. Were they mammoth clusters of stars located at enormous distances, or relatively small clouds of nearby gaseous matter? Until astronomers decided between these possibilities for the luminous spirals, they had no hope of deciphering the meaning in their rapid motions.

A few astronomers held the first view; they argued that the spirals were island universes or true galaxies, enormously large and enormously distant, each containing many billions of stars.[1] In their opinion, the galaxy to which the Sun belonged was only one island universe of stars among many that dotted the vastness of space. But other astronomers preferred the second theory, which held that the luminous spirals were small, nearby objects – little pinwheels of gas, swirling in the space between the stars of the Milky Way.

Some proponents of this view even argued that each spiral was a newborn solar system, with a star forming in the center of the spiral and a family of planets condensing out of the streamers of gas around it.

Hubble settled the controversy. Using the 100-inch telescope, he photographed several spirals with great care, and showed that each one contained enormous numbers of starlike points of light. Hubble found that among these points of light some varied rhythmically in their light output, in the same way as a certain type of star in our Galaxy called a *Cepheid variable star*. He concluded that the points of light were indeed true stars.

---

[1] At the time, these objects were called "spiral nebulas" because no one knew whether or not they were true galaxies. The term "spiral galaxy" came into use later, largely as a result of Hubble's work.

Hubble's observations proved that the spirals were true island universes or galaxies – great clusters of stars, very much like our own Galaxy.

Furthermore, since the spiral galaxies contained so many stars they must be very large; yet their apparent size, as seen in the telescope, was quite small. The implication was that they were extremely far away – far beyond the boundaries of our Galaxy.

Exactly how far away were the spirals? Hubble thought that if he knew the answer to that question, he could solve the mystery of Slipher's retreating galaxies. Hubble at first used a simple method for judging distances called the method of the Standard Candle. This method is used by every person who drives along a narrow road on a dark, moonless night. If a car approaches traveling in the opposite direction, the driver judges how far away it is by the brightness of its headlights. If the lights are bright, the car is close; if they are dim, the car is far away.

Following the same reasoning, Hubble judged the distance to other galaxies by the brightness of the stars they contained. He used the automobile driver's rule of thumb: the fainter the stars in a galaxy, the more distant it is.

An accurate measurement of galactic distances by this method is complicated by the fact that some stars in a galaxy are larger than others and give off greater amounts of light. Hubble therefore used a second method to estimate galactic distances. He observed the Cepheid variable stars in other galaxies. The true brightness of these Cepheids was known from the properties of similar stars in our own Galaxy. This method worked out to distances of about 10 million light-years. Beyond that point, the Cepheid variables in other galaxies were too faint to be seen.

For still greater distances, Hubble developed other methods, such as using the brightness of an entire galaxy as an indication of its distance. Thus, the fainter an entire galaxy, the more distant it is.

In this way, Hubble arrived at values for the distances to about a dozen nearby galaxies. The majority were more than a million light-years away, and the distance to the farthest one was 7 million light-years. These distances were staggering;

they were far greater than the size of our Galaxy, which is 100 000 light-years across. Until Hubble made those measurements, no one knew how big the Universe is.

Next, armed with his list of distance measurements, Hubble turned back to Slipher's values for the speeds of these same galaxies, augmented by his more recent observations of the red shift. He plotted speed against distance on a sheet of graph paper, and arrived at the amazing relationship known as *Hubble's Law: The farther away a galaxy is, the faster it moves.* Hubble's measurements indicated that the relationship was a simple proportion. That is, if one galaxy is twice as far away as another, it will be moving away twice as fast; if it is three times as far, it will be moving away three times as fast; and so on.

Hubble's Law can be simply stated mathematically in the following form; let v be the velocity of recession of a galaxy, and let x be its distance from us. Then

$$v = Hx.$$

H is a constant of proportionality called the Hubble constant. It has units of velocity over distance, and is often expressed in kilometers per second per million light-years. The currently accepted value of the Hubble constant is about 22 kilometers per second per million light-years. A plot of velocity versus distance for a number of galaxies is shown in Figure 2.4. The straight line gives the best average fit to the points plotted on this graph.

2.2.1 Hubble's Law

Hubble's Law of the expanding Universe is one of the great discoveries in science; it is the foundation of the scientific picture of the early Universe. Yet it is a mysterious law. Why should a galaxy recede from us at a higher speed simply because it is farther away?

An analogy will help to make the meaning of this law clear. Consider a lecture hall whose seats are spaced uniformly, so that everyone is separated from their neighbors in front, in back, and to either side by a distance of, say, 3 feet (1 meter). Now suppose the hall expands rapidly, doubling its size in a short time. If you are seated in the middle of the hall, you will

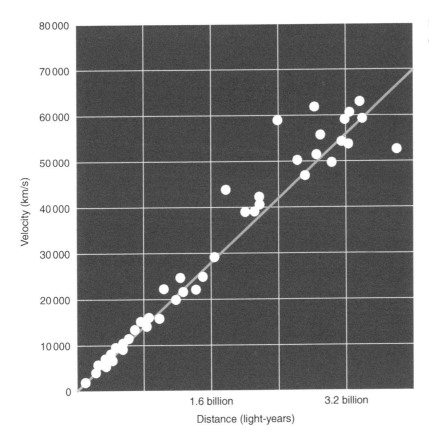

FIGURE 2.4 Velocity versus distance for 46 galaxies.

find that your immediate neighbors have moved away from you and are now at a distance of 6 feet (2 meters). However, a person on the other side of the hall, who was originally at a distance from you of, say, 300 feet (100 meters), is now 600 feet (200 meters) away. In the interval of time in which your close neighbors moved 3 feet farther away, the person on the other side of the hall increased his distance from you by 300 feet (100 meters). Clearly, he is receding at a faster speed.

This is Hubble's Law, or the law of the expanding Universe. It applies not only to the Cosmos, but also to inflating balloons and loaves of bread rising in the oven. All uniformly expanding objects are governed by this law. If the seats in the lecture hall moved apart in any other way, they would pile up in one part of the hall or another; similarly, if galaxies moved outward in accordance with any relationship other than Hubble's Law, they would pile up in one part of the Universe or another.

2.2.2 Is Earth at the center of the expanding Universe?

The description provided by these astronomical discoveries contains a puzzling implication. If all the galaxies in the heavens are moving away from Earth, we must be at the center of the Universe. That notion, commonly held in previous times, was challenged by Copernicus five hundred years ago, and very few people accept it today. Why does modern astronomy lead to a picture of a world that was abandoned by scientists many centuries ago?

The answer is that if you were sitting on a planet in one of the other galaxies in the Universe, you would see the other galaxies around you receding in exactly the same way that an observer in our Galaxy sees our neighbors moving away. Your galaxy would seem to be at the center of the expansion, and so would every other galaxy; but, in fact, there is no center.

To understand this statement more clearly, imagine a very large, unbaked loaf of raisin bread. Each raisin is a galaxy. Now place the unbaked loaf in the oven; as the dough rises, the interior of the loaf expands uniformly, and all the raisins move apart from one another. The loaf of bread is like our expanding Universe. Every raisin sees its neighbors receding from it; every raisin seems to be at the center of the expansion; but there is no center.

2.3 The age of the Universe

When did it happen? When did the Universe explode into being? Knowing the speeds with which the galaxies are moving apart, and how far apart they are at the present time, we can calculate when they were all packed together. The latest measurements of the speeds and distances of the receding galaxies correspond to a value of about 14 billion years for the age of the Universe.

The age of the Universe can also be determined by estimating the ages of the globular clusters in our Galaxy. Globular clusters are compact clusters of stars that are believed to be among the first stars to have formed in the Galaxy (Figure 2.5). Therefore, their ages should give a good indication of the age of the Galaxy. Indirect methods for determining the ages of

FIGURE 2.5 A globular cluster is a spherical cluster that typically contains a few hundred thousand stars. This cluster, called M13, is approximately 23 000 light-years from Earth. The stars in M13 are all crowded within a diameter of only 150 light-years so that the average density of stars in M13 is about 100 times greater than in the neighborhood of the Sun.

S. Down/NOAO

the stars in the globular clusters of our Galaxy indicate that their ages range up to 13 billion years. Assuming that the Milky Way Galaxy (and other galaxies) formed when the Universe was about 1 billion years old, this method also yields an age of about 14 billion years for the Universe.

What is the meaning of 14 billion years? The mind cannot grasp the meaning of such a vast span of time. A million years seems like a very long time, but a billion is a thousand times a million. Nonetheless, nature requires this enormous interval to create its great works. One billion years ago, the Atlantic Ocean did not exist; many of the stars we see in the night sky had not yet been born; and the most advanced form of life on Earth was a simple, multi-celled organism. The face of our planet, the appearance of the heavens, and the shapes of most of the creatures that move across Earth's surface – all these are the product of 1 billion years in the evolution of the Universe.

The mind must stretch its concepts of space and time far beyond their normal limits to comprehend the sweep of the events that make up the history of the Universe. Suppose we adopt a point of view so broad that the tremendous span of a galaxy seems a detail, and the passage of a billion years is like an hour. Imagine the face of a cosmic clock on which one 24-hour day represents the life of the Universe. On this clock,

1 million years is less than a minute, and ten thousand years – the entire span of human civilization – is one-tenth of one second.

Consider the great events in the history of life on Earth within the framework of this analogy. Let the beginning of the Universe occur at midnight; then the galaxies, stars, and planets begin to form 20 minutes after midnight, and continue to form throughout the night and day. At 4 p.m. on the following afternoon, the Sun, Earth, and the Moon appear. At 11:53 p.m., the fishes crawl out of the water; 2 minutes before midnight, the dinosaurs appear; 60 seconds later, they disappear; 1 second before midnight, modern humans appear on the scene.

2.4 The primordial fireball: Proof of the cosmic explosion

The reasoning that leads to the Big Bang theory is indirect; that is, in the astronomer's imagination, the outward motions of the galaxies are projected backward in time until a point is reached at which all galaxies are packed together, and the density and temperature of the Universe are very high. It is as if we saw a cloud of fragments spreading through the air, and inferred from their paths that an explosion had occurred earlier.

In 1948, Ralph Alpher and Robert Hermann, two astrophysicists who were investigating the possibility of a Big Bang, calculated that if a cosmic explosion really did occur, the Universe must have been filled with an intense radiation in the first moments of the explosion. In fact, this radiation would resemble the fireball that forms when a hydrogen bomb explodes. The intensity of the fireball would diminish as the Universe expanded, but a small remnant of the primordial fireball radiation should still be present in the Universe today. Furthermore, it should be detectable with sensitive radio telescopes.

In 1965, Arno Penzias and Robert Wilson of the Bell Laboratories discovered that Earth is bathed in a faint glow of radiation coming from every direction in the heavens. Their measurements revealed that the puzzling radiation arrived at Earth with equal intensity from all parts of the sky. It did not

FIGURE 2.6 Arno Penzias (right) and Robert Wilson, discoverers of the primordial fireball radiation, in front of the horn antenna used in their epochal research. Lucent Technologies, Inc

seem to come from the direction of the Sun, a galaxy, or any other particular object in the sky; the entire Universe seemed to be the source. This was just the characteristic expected for the cosmic fireball radiation.

Penzias and Wilson were not looking for proof of the Big Bang cosmology when they made this observation; they were measuring the intensity of radio waves in a large antenna that had been set up for a completely different purpose (Figure 2.6). They were unable to explain the source of the uniform radiation until a colleague told them about a lecture he had heard on the possibility of finding radiation left over from the fireball that filled the Universe at the beginning of its existence. Then they realized they had detected the cosmic fireball.

Radio astronomers have made many measurements on the fireball radiation since it was first discovered. The measurements show that the spectrum of the radiation – that is, the change in its intensity with wavelength – very closely matches the spectrum of the radiation emitted by a heated object (Figure 2.7). This characteristic pattern of wavelengths is called a *black-body spectrum*. The same spectrum is also observed for the radiation produced by the hot gases resulting from an explosion. The pattern of wavelengths measured by the radio astronomers is the fingerprint of the Big Bang. The agreement

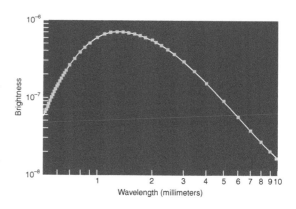

FIGURE 2.7 Spectrum of the primordial fireball radiation measured by the Cosmic Background Explorer (COBE) satellite. The observed data points fall almost exactly along the curve of radiation for a gas at a temperature of 2.726 degrees above absolute zero. Shortly after the Big Bang the temperature of the fireball radiation was much higher, but it decreased with time as the Universe expanded.

between the spectrum of the Penzias–Wilson radiation and a black-body spectrum has convinced nearly all astronomers that the Big Bang really did occur.

2.5 Conditions in the evolving Universe

Why did the Universe explode 14 billion years ago? The Big-Bang cosmology does not answer this question, nor does it provide information on conditions in the Universe prior to the cosmic explosion. However, accepting first that the Universe was once in an exceedingly dense, hot state, cosmologists can predict the major events in the subsequent evolution of the Universe with some confidence.

In the first moments after the Big Bang, the density of matter and energy in the Universe was very great, and the temperature ranged up to trillions of degrees. At about one-millionth of a second after the Big Bang, the Universe cooled down enough so that the basic particles of matter – protons, neutrons, and electrons – became stable. Protons represent the nuclei of hydrogen atoms, the simplest element. After a few minutes, the temperature decreased to about a billion degrees, and protons and neutrons began to stick together in groups of four to form atomic nuclei of the element helium. Calculations suggest that about 25% of the matter in the Universe was transformed into helium in the first few minutes after the Big Bang. This helium is called *primordial helium.*

It might be expected that once helium formed, other heavier elements would then be built up until the whole periodic table of the elements existed. However, calculations on the early history of the Universe indicate that this did not occur. The principal reason is that a nucleus with five neutrons and protons, which would be the natural successor to helium in a chain of nuclear reactions, does not exist. The next stable nucleus after helium is lithium, which has a total of six neutrons and protons. The gap between a four-particle nucleus and a six-particle nucleus turns out to be an impossible one to cross under the conditions of rapidly falling temperature and density that existed in the first minutes after the Big Bang.[2] Thus, the synthesis of heavier elements beyond helium ceased when the Universe was about 30 minutes old, and was not resumed until a billion years later, when star formation began.

After helium formed, the expansion of the Universe continued and its temperature continued to drop. The matter of the Universe existed at that time mainly in the form of hydrogen nuclei (i.e., protons), helium nuclei, and free electrons. Atoms did not yet exist, because whenever an electron was captured into an orbit around a nucleus to form an atom, it was knocked out of its orbit almost immediately, under the smashing impact of the violent collisions that occur at such high temperatures.

However, when the Universe was a few hundred thousand years old and the temperature had dropped to 3000 °C, collisions were no longer violent enough to dislodge electrons from their orbits. From that time on, the hydrogen and helium in the Universe were present mainly in the form of neutral atoms. The oldest stars and gas clouds are composed of about 25% helium.

With the passage of time, the gaseous clouds of primordial hydrogen and helium cooled and condensed into galaxies and, within the galaxies, into stars. The formation of these first stars composed entirely of hydrogen and helium probably began shortly after the formation of the first galaxies, when the Universe was between a few hundred million and a billion years old. As we will see in Chapter 3, the synthesis of

[2] A small amount of heavy hydrogen, or deuterium, with one proton and one neutron in its nucleus was also produced in the earliest moments of the Big Bang. Traces of the elements lithium (three protons and three or four neutrons) and beryllium (four protons and three neutrons) were also formed.

elements heavier than helium resumed in the interiors of stars. After about 14 billion years of continuing expansion, star formation, and heavy-element production, the Universe reached the state in which it exists today.

2.6 The horizon problem

The successes of the Big Bang cosmology raise some interesting questions. Suppose we look out in the Universe in one particular direction and see a galaxy at a distance of, say, 10 billion light-years. Then we turn and, looking in the opposite direction, we see another galaxy, also at a distance of 10 billion light-years. The two galaxies are 20 billion light-years apart (or were that far apart when the light by which we observe them first set out on its way to us).

Since the Universe is only 14 billion years old, a light ray that started out from one of these galaxies 14 billion years ago, and has been traveling steadily across the Universe toward the other galaxy, is still en route. Because of the limited age of the Universe, it has only covered 14 billion light-years thus far.

This radiation will not reach the other galaxy until another 5 billion years has elapsed. Yet observations show that the intensity of the primordial fireball radiation is the same in those two opposite corners of the Universe, to better than one part in 10 000. That implies that these distant regions of the Universe must have already exchanged energy and come to the same temperature. How can that be, if neither matter nor radiation can yet have passed between them?

This is called the *horizon problem*. From one side of the observable Universe, the other side is over the horizon and inaccessible. Yet astronomical observations suggest that the two sides of the Universe were once in contact.

2.7 The flat Universe problem

Although the Universe is expanding today, it may not continue to expand into the indefinite future. Gravity, pulling back on the outward-moving galaxies, slows their rate of expansion. If the pull of gravity is sufficiently great, it can bring the expansion to a halt. When that happens, the components of the Universe stand momentarily at rest, and then, drawn together by their mutual gravitational attraction, they commence to move toward one another. Slowly at first, and then with increasing momentum, the Universe collapses under the pull of gravity. Soon the galaxies rush toward each other with an inward movement as violent as the outward movement of the Big Bang. After sufficient time, they come into contact, the galaxies mix and are heated by compression, and the Universe returns to the heat and chaos from which it emerged billions of years ago.

On the other hand, if the pull of gravity is not sufficiently strong, the Universe will continue to expand forever, even though the expansion is slowed down to some degree by gravity. As the expansion continues, eventually the stars burn out, one by one, until all ends in darkness.

What is the fate of the Universe? Will it come to a halt and collapse, or expand forever? The answer depends on the strength of the gravitational force that pulls the components of the Universe toward one another. That force, in turn, depends on the amount of matter in the Universe – or, more accurately, on the density of matter in the Universe. A high density of matter means that on the average, the masses that populate the Universe are relatively close to one another and, therefore, the force of gravity that attracts each to its neighbor is relatively strong. A low density of matter means that the masses in the Universe are relatively far from their neighbors, and the gravitational attraction is correspondingly weak.

Calculations show that there is a critical threshold density that determines the fate of the Universe. If the density of matter is less than the critical density, the Universe will expand forever. If the density of matter is greater than the critical density, the Universe will come to a halt and collapse. What do the observations reveal? When astronomers add up all the matter that we can see in the Universe in a visible form, as luminous stars and galaxies, the resultant density of matter turns out to be only one-hundredth of the critical density needed to halt the expansion. Thus, it would appear that the Universe will expand forever.

FIGURE 2.8 Astronomers analyzed the distribution of galaxies in the giant galactic cluster CL0024+1654 (4.5 billion light-years from Earth). They used this information to map the distribution of otherwise invisible dark matter (shown in blue) associated with the massive cluster.

Jean-Paul Kneib (Laboratiore d' Astrophysique de Marseille, France), European Space Agency and NASA

However, this is not the last word because discoveries in recent years have revealed that the contents of the Universe include more than the matter we can see. The Universe also contains a large amount of *dark matter*, which reveals itself only by its influence on the motions of the stars and galaxies (Figure 2.8). According to the latest observations, the Universe contains 30 times more dark matter than visible matter. This means that the density of matter in the Universe is still about a third of the critical density needed to halt the expansion. So, even with dark matter included, the best guess at the present time is still that the Universe will expand forever.

The reason this conclusion leads to a problem becomes clear when we extend our inquiries back in time. The calculations show that while the density of matter today is smaller than the critical density by a factor of three, as we go back in time toward the Big Bang the difference between the two densities diminishes. Shortly after the Big Bang, the density of matter in the Universe was extremely close to the critical density. In fact, one second after the Big Bang it differed from the critical density by less than one part in a trillion.

Furthermore, the calculations yield the surprising result that if the density at that early time had differed from the critical density by as much as one part in a million, the Universe would have turned out to be completely different than it is today. Suppose, for example, that the density one second after the Big Bang had been less than the critical density by one part in a million, then the elements of matter in the Universe would have flown apart too rapidly for galaxies, stars, and planets to form. That means we would not be here today.

Suppose, on the other hand, that the density of matter at that early time had been greater than the critical density by one part in a million; then the expanding Universe would have come to a halt and collapsed in on itself too soon for life to evolve on any planets that eventually formed. Again, we would not be here.

So, unless the density in the early Universe was extremely close to the critical density, a universe containing stars, planets, and life could not have come into being.

That is the essence of the problem. Why should the density of matter have had just that particular value – the so-called critical density – that is so favorable for the emergence of a universe containing living beings? The density of the early Universe could just as well have been 10 times greater than the critical density, or 1000 times less. There is no present explanation in the Big Bang theory for the seemingly fortuitous fact that the density of matter has just the right value for the evolution of a benign, life-supporting universe.

This puzzling set of ideas is called the *flatness problem* in cosmology because of the language of curved space used in Einstein's theory of general relativity. According to the equations of general relativity, space is curved by the presence of matter. When cosmological theories are expressed within the framework of general relativity, it turns out that in a universe in which the density of matter is greater than the critical density, space has a positive curvature, analogous to the curvature of the surface of a sphere.

A universe of this kind is said to be "closed" because, in such a universe, a person who travels in a straight line for billions of years will eventually, owing to the curvature of space, come back to his or her starting point.

In a universe in which the density of matter is less than the critical density, space has a negative curvature, analogous to the curvature of a saddle. Such universes are said to be "open" because a straight-line journey across such a universe will never bring the traveler back to his or her starting point.

But if the density in a universe is precisely equal to the critical density, space has zero curvature, and is said to be "flat." This is approximately the case for the Universe we inhabit; its space is flat, or nearly so.

Yet, as noted, there is no reason for the space in the Universe to be so flat, either within the context of the Big Bang theory of in the context of physics as a whole, at least as we currently understand it. Thus, the questions relating to the apparently fortuitous agreement between the density of matter and the critical density can be restated as: Why is space so flat?

This is the flatness problem. It achieves a special significance since a flat or very nearly flat universe appears to be the only kind in which life could have evolved. The fact that the Universe is so close to being flat today, and was even closer in an early epoch, leads some cosmologists to believe that the flatness of our Universe is not a fortuitous accident but a requirement of some basic physical law or phenomenon.[3]

2.8 The inflationary Universe

One possible solution to the flatness problem is the idea of an inflationary early Universe, first proposed by astrophysicist

---

[3] Flatness is not the only case in which the properties of the Universe seem within extremely narrow limits to be suitable for life. For example, if we imagine that the force acting between nuclear particles – neutrons and protons – is a little bit stronger than we know it actually to be, we expect to find that in the early moments of the Universe, a little more helium was made, and later in stars, a little more carbon.

The astronomers have found nothing of the sort. Their calculations indicate that if the strength of the nuclear force is increased by roughly one percent over its present value, all hydrogen is gobbled up into helium by nuclear reactions, and the stars, now made of helium, live for such a short time that no life can possibly have a chance to evolve around them. And if the strength of the nuclear force is decreased by about one percent, nuclear reactions take place so slowly in the Universe that none of the elements on which life depends – carbon, nitrogen, oxygen, and so on – can be formed in the 14 billion years available since the Big Bang. The evidence that the Universe is somehow designed for life is called the *Anthropic Principle*.

Alan Guth. Guth's proposal goes back to the very earliest moments of the Universe, shortly after the Big Bang. At this time, he suggests, the Universe, cooling very rapidly as it expanded, entered a super-cooled state analogous to the super-cooled state of water. In this state, water remains a liquid even if its temperature has dropped below freezing. When the super-cooled water finally solidifies to ice, an enormous pulse of energy in the form of heat is released in a very short time. Just so, in the super-cooled Universe, when condensation occurs, the Universe receives a tremendous jolt of energy in a very short time.

The energy leads to an enormous increase in the rate of expansion of the Universe, and a period of rapid inflation of the Universe sets in. According to Guth's theory, at the beginning of the period of inflation, the entire observable Universe was considerably smaller than an atomic nucleus. During the period of inflation, its size increased with explosive rapidity. After 10^{-32} seconds, it was about 30 feet (10 meters) across. At approximately this time, the inflationary period ended, and the Universe returned to its "normal" rate of expansion.

The key point in the theory is that prior to the inflationary period, all the elements of matter in the present Universe were close enough to exchange energy and reach a uniform temperature.

Thus, the inflationary hypothesis seems to solve the horizon problem. It is no longer difficult to understand why two regions at opposite ends of the observable Universe would be at the same temperature. Inflation also solves the flatness problem because, regardless of whether the curvature of the early Universe was positive or negative, at the end of the inflationary period of extremely rapid expansion the part of the Universe we can see would be essentially flat, just as a patch of the curved surface of a sphere would appear essentially flat after an enormous increase in the sphere's radius.

The details of inflation may be wrong – it is hard to be confident of our ability to describe the state of the Universe 10^{-32} seconds after the Big Bang – but the theory provides a possible solution to the vexing problems of flatness and the horizon.

2.9 Dark energy and an accelerating Universe

We noted earlier that as the Universe expands and the galaxies move farther apart, the attractive force of gravity pulls back on the retreating galaxies and tends to slow down the expansion. However, recent evidence indicates that instead of slowing down, the expansion of the Universe is actually accelerating. The explanation of this result may be connected with Einstein's equations of general relativity, which have a term that can act like a repulsive force pervading the Universe and driving the galaxies apart.

The evidence for an accelerating Universe involves a type of exploding star or supernova called a Type Ia supernova. All supernovas of this type have approximately the same intrinsic brightness. The apparent brightness of these Type Ia supernovas, as observed by us, depends on their distance; the greater the distance, the dimmer the supernova will appear to be.

Now suppose a Type Ia supernova explodes in a galaxy several billion light-years from us. If the pull of gravity were slowing the expansion of the Universe, the distance from us to that supernova would be less than if the Universe had been expanding at a constant rate. Thus, the supernova would appear to be brighter than it would be if the rate of expansion of the Universe were constant.

Observations of the apparent brightness of several Type Ia supernovas in distant galaxies have yielded the unexpected result that these supernovas are, in fact, *dimmer* than we would estimate assuming a constant rate of expansion of the Universe. Consequently, the distances to these galaxies must be *greater* than they would be in a Universe expanding at a constant rate. Instead of slowing down, the expanding Universe must actually be *accelerating*.

What could cause such an acceleration in the expansion of the Universe? Astronomers think that the galaxies are moving apart at a faster and faster rate because of a repulsive force they call *dark energy*, which may be an inherent property of the vacuum that pervades all space at the smallest scale. This dark energy provides a negative pressure that causes the expansion to accelerate. In this latest picture of the Cosmos, most of the Universe is composed of dark matter and this dark energy (Figure 2.9).

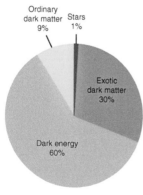

FIGURE 2.9 Visible matter in stars makes up less than 1% of the contents of the Universe. Dark matter and dark energy make up most of the remainder.

If this result stands the test of time and additional observation, it will enable us to predict the future of the Universe. The galaxies of the Universe fly apart forever; within each galaxy, the old stars and short-lived massive stars burn out, one by one; fewer and fewer new stars are born to replace them as the density of the primordial gases diminishes; eventually the entire Universe grows dark.

2.10 Summary

By the end of this chapter you should be familiar with the following concepts and topics:

The origin and evolution of the Universe
Evidence for the explosive origin of the Universe
How the speeds of galaxies are measured: The Doppler shift
The expansion of the Universe
Hubble's Law
Is Earth at the center of the expanding Universe?
The age of the Universe
Favorable density of matter for a universe containing living
   beings
The fate of the Universe and its contents

Questions

2.1. How did Slipher (and later Hubble) determine that many galaxies were moving away from Earth?

2.2. What is the best evidence that the Universe began in an explosion?

2.3. Explain what is meant by the statement, "When you look out into space you look back in time."

2.4. It is believed that a massive black hole (of about 1 million solar masses) exists at the center of our Galaxy. What would be the radius of its event horizon?

CHAPTER 3

Life from the cosmic cauldron

Life history of a star; Creation of the chemical elements within stars

Far from being detached observers, we are an intrinsic part of our Universe. The materials that make up our bodies and our planet Earth, and the proportions in which they occur, are the result of events that took place in other stars in our Galaxy in times before the Solar System even existed. Approximately 14 billion years ago, the Universe began in a dramatic cosmic explosion and the primordial elements hydrogen and helium were formed. Every other element in existence today has been synthesized out of hydrogen and helium in a star at some point during the history of the Universe. Today the Universe contains over 90 elements, some of which were crucial for the formation of planets like Earth, and for the existence of life.

All stars begin in the same way, yet their ultimate fates can be very different. In this chapter we will look at the life cycle of a star, and see how everything in our Universe owes its existence to the processes that occur as stars evolve. Beginning with star formation, we will follow the series of events throughout a star's life, learning how the elements now contained in our bodies were formed, and why lighter elements are more abundant in the Universe than heavier ones. The fate of a star depends on its mass, and we will see how massive stars eventually explode in supernova events, creating heavy elements such as silver and gold in their final minutes. We also look at astronomical observations that indicate the existence of neutron stars, black holes, and quasars – objects all resulting from the explosions of stars.

3.1 Birth of a star

When the Universe was very young, shortly after it exploded into being, it contained only the elements hydrogen and helium. There were no other elements – no iron, oxygen, and so on – and there were no stars and no planets. After the Universe was about 1 billion years old, stars began to condense out of the primordial hydrogen and helium. They continued to form as the Universe aged. Our star – the Sun – arose in this way 4.6 billion years ago, when the Universe was more than 9 billion years old. Many stars came into being before the Sun was formed; many others formed after the Sun appeared. This process continues, and through telescopes we can see stars forming today, out of condensed pockets of gas in outer space.

How does a star come into being? The answer lies in the action of gravity on the clouds of hydrogen in space. The Universe is filled with thin clouds of hydrogen, which surge and eddy in the space between the stars. In the swirling motions of these tenuous clouds, atoms sometimes come together to form small pockets of gas (Figure 3.1). These pockets are temporary condensations in an otherwise highly rarefied medium. Normally the atoms fly apart again in a short time as a consequence of their random motions, and

Previous page: A region of star formation in our Galaxy.
  Royal Observatory, Edinburgh

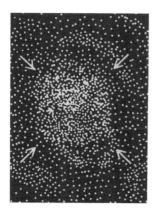

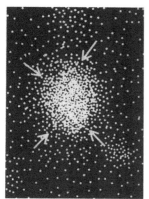

FIGURE 3.1 A cloud pulled together by gravity. It is believed that stars are born in swirling clouds of hydrogen gas. If the atoms in one region of such a cloud come together by accident or are forced together by the pressure of the surrounding clouds, the force of gravity pulls the atoms still closer together, forming a condensed pocket of gas (top). The continuing action of gravity compresses the pocket of gas (bottom).

the pocket of gas quickly disperses into space. However, each atom exerts a small gravitational attraction on its neighbors, which counters the tendency of the atoms to fly apart. If the number of atoms in the pocket of gas is large enough, the accumulation of all these separate forces will hold it together indefinitely. It is then an independent cloud of gas, preserved by the gravitational attraction of each atom in the cloud to its neighbors (Figure 3.2).

With the passage of time, the continuing influence of gravity, pulling all the atoms closer together, causes the cloud to contract. The individual atoms "fall" toward the center of the cloud under the force of gravity; as they fall, they pick up speed and their energy increases. Collisions among these atoms heat the gas and raise its temperature. The shrinking, continuously self-heating ball of gas is an embryonic star.

Photographs taken in certain regions of the sky show small, dark globules of matter that look like stars, or clusters of stars, in the process of formation (Figure 3.3). The globules are dark because they contain substantial amounts of dust in addition to gaseous material. The gas is relatively transparent, but the grains of dust absorb light very effectively.

As the gas cloud contracts under the pressure of its own weight, the collisions of the gas molecules grow more violent and the temperature at the center mounts steadily. When it reaches about 10 000 degrees Celsius, the hydrogen atoms in the gas collide with sufficient violence to dislodge all electrons from their orbits around the protons.

At this stage, the globe of gas has contracted from its original size, which was about 10 trillion miles (16 trillion kilometers) in diameter, to a diameter of about 100 million miles (160 million kilometers). To understand the extent of the contraction, imagine a battleship shrinking to the size of a grain of sand.

The huge ball of gas – now composed of separate protons and electrons – continues to contract under the force of its own weight, and the temperature at the center rises further. After 10 million years, the temperature has risen to the critical value of 10 million degrees Celsius. At this time, the diameter of the ball has shrunk to about 1 million miles (1.6 million kilometers), which is about the size of our Sun and other typical stars.

FIGURE 3.2 Hubble Space Telescope image of the Eagle Nebula, a nearby star-forming region 7000 light-years from Earth. The image shows newborn stars emerging from dense, compact pockets of interstellar gas and dust.

Paul Scowen and Jeff Hester, School of Earth & Space Exploration, Arizona State University

FIGURE 3.3 Bok globules (named after astronomer Bart Bok) are small dark clouds made of gas and dust that are in the process of condensing to form stars. These Bok globules are found in a region of glowing hydrogen gas.

© Anglo-Australian Observatory/David Malin Images

Why is 10 million degrees Celsius a critical temperature? The explanation is connected with the forces between the protons in the contracting cloud. When two protons are separated by large distances, they repel one another electrically because each proton carries a positive electric charge. But if the protons approach within a very close distance of each other, the electrical repulsion gives way to the even stronger force of nuclear attraction. The protons must be closer together than one ten-trillionth of an inch for the nuclear force to be effective. Under ordinary circumstances, the electrical repulsion serves as a barrier to prevent as close an approach as this. In a collision of exceptional violence, however, the protons may pierce the electrical barrier that separates them, and come within the range of their nuclear attraction. Collisions of the required degree of violence first begin to occur when the temperature of the gas reaches 10 million degrees Celsius.

Once the barrier between two protons is pierced in a collision, they pick up speed as a result of their nuclear attraction and rush rapidly toward each other. In the final moment of the collision the force of nuclear attraction is so strong that it fuses the protons together into a single nucleus. At the same time the energy of their collision is released in the form of heat and light. This release of energy marks the birth of the star (Figure 3.4).

The energy passes to the surface and is radiated away in the form of light, by which we see the star in the sky. The energy release, which is a million times greater per pound than that produced in a TNT explosion, halts the further contraction of the star, which lives out most of the rest of its life in a balance between the outward pressures generated by the release of nuclear energy at its center and the inward pressures created by the force of gravity.

If the contracting cloud contains less than 10% of the Sun's mass (about 100 times the mass of a giant planet like Jupiter), then its internal temperature cannot reach the level required to initiate the fusion of two protons. This object will never become a star. The result is a *brown dwarf* – a small, dim object intermediate in size between a star and a planet.

The first brown dwarf was discovered in 1995, and today several dozen are known. Although these dim objects are

FIGURE 3.4 The Orion
Nebula, a birthplace of stars.
The nebula lies about 1500
light-years from Earth and
has a mass of about 300 solar
masses.
   NASA, ESA, M. Robberto
(Space Telescope Science
Institute/ESA) and the
Hubble Space Telescope
Orion Treasury Project Team

difficult to find, astronomers estimate that they are as numerous as stars.

The fusion of two protons is only the first step in a series of
reactions by which nuclear energy is released during the life
of the star. In subsequent collisions, two additional protons
are joined to the first two to form a nucleus containing four
particles. Two of the protons shed their positive charges to
become neutrons in the course of the process. The result is a
nucleus with two protons and two neutrons. This is the
nucleus of the helium atom. Thus, the sequence of reactions
transforms protons, or hydrogen nuclei, into helium nuclei.

3.2 Stellar evolution

The fusion of hydrogen into helium is the first and longest stage
in the history of a star, occupying about 90% of its lifetime.
Throughout this long period of the star's life its appearance
changes very little, but toward the end of the hydrogen-
burning stage, when most of the hydrogen has been converted
into helium, the star begins to show the first signs of age. The

FIGURE 3.5 Hubble Space Telescope image of the star Betelgeuse (left) , the upper right-hand star in the constellation Orion (right), a red giant star about 510 light-years from Earth.

Andrea Dupree (Harvard-Smithsonian CFA), Ronald Gilliland (Space Telescope Science Institute, STScl), NASA and ESA

center of the star is now nearly pure helium, which has accumulated there as the "ashes" from the burning of hydrogen. This large lump or core of helium at the center leads to a swelling and reddening of the star's outer layers, which commences imperceptibly and progresses until the star has grown to a huge red ball 100 times larger than its original size.

The Sun will reach this stage in another 6 billion years, at which time it will have swollen into a vast sphere of gas engulfing the planets Mercury and Venus and reaching out nearly to the orbit of Earth. This red globe will cover most of the sky when viewed from our planet. Unfortunately, we will not be able to linger and observe the magnificent sight, because the rays of the swollen Sun will heat the surface of Earth to 2000 °C and eventually evaporate its substance. Perhaps one of the moons of the outer planets will be a suitable habitat for us by then. More likely, we will have fled to another part of the Galaxy.

Such distended, reddish aging stars are called *red giants* by astronomers. An example of a red giant is Betelgeuse, a bright star in the constellation Orion, which appears distinctly red to the naked eye (Figure 3.5). Another is the star called V838 Monocerotis. In January 2002, this star showed three unusual bursts of light. Astronomers concluded that the surges of energy occurred as the expanding red giant swallowed three of its orbiting planets one by one.

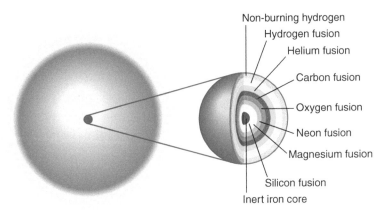

Non-burning hydrogen
Hydrogen fusion
Helium fusion
Carbon fusion
Oxygen fusion
Neon fusion
Magnesium fusion
Silicon fusion
Inert iron core

FIGURE 3.6 The structure of a star late in life. At the beginning of a star's life it burns hydrogen to make helium. In the later stages of its life, helium burns to make carbon. In a large star, light elements are combined into heavier and heavier elements up to iron. A cross-section of a star shows concentric layers in which the production of the heavier elements is taking place.

A star continues to live as a red giant until its reserves of hydrogen fuel are exhausted. With its fuel gone, the red giant can no longer generate the pressures needed to maintain itself against the crushing inward force of its own gravity, and the outer layers begin to fall in toward the center. The red giant collapses.

At the center of the collapsing star is a core of pure helium, produced by the fusion of hydrogen throughout the star's earlier existence. Helium does not fuse into heavier nuclei at the ordinary stellar temperature of 10 million degrees Celsius, because the helium nucleus, with *two* protons, carries a double charge of positive electricity, and, as a consequence, the electrical repulsion between two helium nuclei is stronger than the repulsion between two protons. A temperature of 100 million degrees is required to produce collisions sufficiently violent to pierce the electrical barrier between helium nuclei.

As the star collapses, however, heat is liberated and its temperature rises. Eventually the temperature at the center reaches the critical value of 100 million degrees. At that point, helium nuclei begin to fuse in groups of three to form carbon nuclei, releasing nuclear energy in the process and rekindling the fire at the center of the star. The additional release of energy halts the gravitational collapse of the star. It has obtained a new lease of life by burning helium nuclei to produce carbon (Figure 3.6).

In stars the size of the Sun, the helium-burning stage lasts for about 100 million years. At the end of that time the reserves of fuel, composed now of helium rather than hydrogen, once again are exhausted, and the center of the star is filled with a residue of carbon nuclei. These nuclei, possessing *six* positive electrical charges, are separated by an even-more formidable electrical barrier than helium nuclei, and collisions of even greater violence are required for its penetration. The 100-million-degree temperatures that fuse helium nuclei are not adequate for the fusion of carbon nuclei; no less than 500 million degrees Celsius is required.

Because the temperatures prevailing within the red giant fall short of 500 million degrees, the nuclear fires die down as the carbon accumulates, and the star, once again lacking the resources needed to sustain it against the weight of its outer layers, begins to collapse a second time under the force of gravity.

All stars lead similar lives up to this point, but their subsequent evolution and manner of dying depend on their size and mass. The small stars shrivel up and fade away, while the large ones disappear in a gigantic explosion. The Sun lies well below the dividing line; we are certain that it will fade away at the end of its life.

3.3 White dwarfs and supernovas

The paths of the small and large stars diverge because of differences in the amount of heat generated during the second collapse, at the end of the red giant stage. In a small star, the collapse generates a modest amount of heat, and the temperature at the center fails to reach the 500-million-degree level required for the ignition of carbon nuclei. Thus, the nuclear fire is never rekindled. Instead the collapse continues, until finally the matter within the star is so compressed that it resists a further reduction in size. The star then remains forever in this highly compressed state. Roughly the size of Earth, it has been squeezed by the force of its own weight into a space only a millionth as large as the volume it originally occupied. A teaspoon of matter from the center of this compact body would weigh 10 tons (10 000 kilograms).

FIGURE 3.7 The planetary nebula NGC 7293. Gases ejected from a red giant star in the final stages of its life. (The name "planetary nebula" comes from the fact that astronomers who first observed these objects thought that they resembled solar systems in the process of formation.) The star that ejected these gases is the center of the spherical glowing shell. This nebula is 450 light-years from Earth.

© Anglo-Australian Observatory/David Malin Images

Although the center of the star never gets hot enough to burn carbon, the temperature of the surface rises sufficiently that the star appears white-hot to the eye. These stars are surrounded by a spherical shell of gases ejected from the red giant star. These shrunken white-hot stars are called *white dwarfs*. Slowly the white dwarf radiates into space the last of its heat. In the end its temperature drops, and it fades into a blackened corpse (Figure 3.7).

A very different fate awaits a large, massive star. The largest stars are about 100 times more massive than our Sun. For stars more than eight times the mass of the Sun, the weight of the star is so great its collapse generates an enormous amount of heat, greater than the heat generated in the creation of the white dwarf. Soon the temperature reaches the critical level of 500 million degrees at which carbon nuclei fuse together. The fusion of the carbon nuclei forms still heavier elements ranging from oxygen up to sodium.

Eventually, the carbon fuel reserves are also exhausted; once again their exhaustion is followed by further stages of collapse, heating, and renewed nuclear burning, leading to the production of still heavier elements.

In this way, through the alternation of collapse and nuclear burning, a massive star successively manufactures all elements up to iron. But iron is a very special element. This metal, which lies halfway between the lightest and heaviest elements, has an exceptionally compact nucleus, whose neutrons and protons are so tightly packed that no energy can be squeezed from it in any sort of nuclear reaction. In fact, nuclear reactions in iron absorb energy; they have the same effect as cold water thrown on hot coals. When a large amount of iron accumulates at the center of the star, the fire cannot be rekindled; it goes out for the last time, and the star commences a final collapse under the force of its own weight.

The ultimate collapse is a catastrophic event. The iron nuclei at the center soak up the energy of the star as fast as it is generated, and the collapsing materials, meeting negligible resistance, fall in toward the center at enormous speeds, covering a million miles in less than a minute. They pile up around the material at the center in a dense lump, squeezing the core of the star, and creating enormous pressures. When the pressure at the center is sufficiently great, the collapse comes to a halt. The collapsed star, compressed like a spring, is momentarily still – and then it rebounds in a violent explosion, leaving nothing behind but the squeezed remnant of the star's original core. Sometimes the star blows up as soon as its carbon ignites, without going through the sequence of events leading up to the formation of iron. The end result is the same: a violent explosion occurs; the star is destroyed; and its materials are dispersed to space.

The explosion generates temperatures ranging up to trillions of degrees. At these temperatures some of the nuclei in the exploding star are disintegrated, and many neutrons are freed. The neutrons are captured by other nuclei, building up the heavier elements, such as lead, gold, and uranium. In this way the remaining elements of the periodic table, extending beyond iron, are manufactured in the final moments of the star's life. This is why these heavy elements, including silver, gold, platinum, and so on, are so rare in the Universe and so rare on Earth. The small amounts that exist in the Universe were manufactured in minutes or even seconds, in the last gasp of a star's life. Elements like carbon and oxygen are more abundant because they were manufactured steadily within the star over the course of many millions of years.

FIGURE 3.8 A dying star. One of the earliest reported supernovas was observed by Chinese astronomers in AD 1054. Today at the position of this supernova there is a large cloud of gas known as the Crab Nebula, shown in the photograph, which is expanding rapidly.

European Southern Observatory

The explosion of the star hurls out to space all the elements it has been manufacturing during its lifetime. The entire episode lasts a few minutes, from the onset of the collapse to the final explosion. This is a short interval for the demise of an object that may have lived for 10 million years.

The exploding star is called a *supernova*. Supernovas blaze up with a brilliance many billions of times greater than the brightness of the Sun. If the supernova happens to be nearby in our Galaxy, it appears suddenly as a new star in the sky, brighter than any other, and easily visible with the naked eye in the daytime. One of the earliest reported supernovas was a brilliant explosion recorded by Chinese astronomers in AD 1054. At the position of this supernova there is today a great cloud of gas known as the Crab Nebula, expanding at a speed of about a thousand miles (1500 kilometers) per second, which contains the remains of the star that exploded over 900 years ago (Figure 3.8).

3.4 Abundances of the elements in the Universe and the life cycle of stars

The story of the substances of the Universe starts with the primordial elements hydrogen and helium. These two elements, which were present in the Universe from the early moments of its existence, are by far the most abundant elements. In terms of number of atoms, hydrogen and helium make up roughly 94% and 6% of all matter. In terms of mass, they make up 75% and 25%, respectively. The remaining atoms make up a few in a thousand of the total number of atoms. Every atom of this material has been synthesized out of hydrogen and helium in some star at some point during the history of the Universe.

For the student of Earth and the planets, the important lesson to be drawn from this astronomical story of the birth and death of stars is the fact that all the chemical elements heavier than hydrogen and helium were manufactured in other stars before the Sun and Earth were formed.

This means that the relative abundances of these elements – including carbon, nitrogen, oxygen, silicon, iron, and all the other substances important to Earth and its inhabitants – are present in the Universe in amounts that depend, first, on the number of stars that have lived and died since the Universe came into existence and second, on the laws of nuclear reactions under the high-temperature conditions prevailing at the center of a star. The accompanying chart (Figure 3.9) shows the relative abundances of all the elements in the Universe (determined mainly from studies of the Sun and other stars) plotted against their atomic number, ranging from 1 for hydrogen to 92 for uranium.

After the primordial hydrogen and helium, the next most abundant elements are carbon, nitrogen, and oxygen. These are very abundant because they are the first elements to be manufactured in stars out of the primordial substances and, also, carbon and oxygen, in particular, have exceptionally stable nuclei. That is, in nuclear reactions the formation of these stable units tends to be favored.

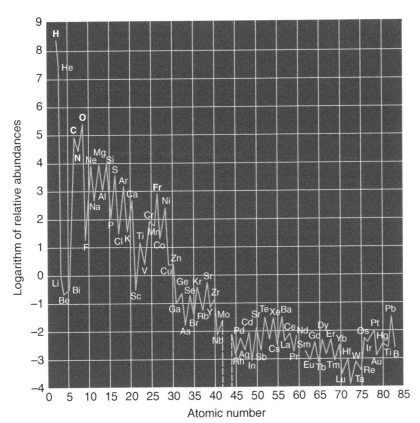

FIGURE 3.9 Abundance of elements in the Universe. The abundances are plotted on a logarithmic scale. Note the drop off in abundance after iron (Fe) and nickel (Ni) with atomic numbers 26 and 28.

After carbon, nitrogen, and oxygen, the next most abundant group runs from neon to calcium. In this group are many of the substances that make up rocklike materials and the body of Earth. There is a general downward trend in the abundances of these elements, with increasing mass and atomic number. The reason for this downward trend is that the elements in question are formed in smaller and smaller regions of the stellar cores, and at successively later stages in the lifetime of a star. As the star's life proceeds, the rate of its evolution speeds up, so that less time is available for the manufacture of an element formed later. Therefore, a smaller amount of the element is manufactured.

Superimposed on the general downward trend in abundance are large variations up and down from element to element. These variations, which give the graph its jagged appearance, result from differences in the stability of a nucleus; the most stable nuclei are more likely to be formed and their abundance in the Universe is greater.

The general downward trend with increasing atomic number is interrupted when the graph reaches iron. Iron is more abundant than the somewhat lighter elements because it is the most stable nucleus in the Universe. For this reason, when the life of a star reaches the stage in which iron can be manufactured at the center of a star, a very large amount of iron is produced. The same is true to a somewhat lesser extent for nickel, another heavy element with a relatively stable nucleus.

After iron, the abundances of the elements fall off sharply by about a factor of 10 000. The explanation for this rapid drop is that the appearance of large amounts of iron in the center of a star triggers the violent collapse and subsequent supernova explosion that mark the end of a massive star's life. Very little time is available for the elements beyond iron to be manufactured, once this explosion has occurred. All of them, from copper and zinc up to lead and uranium, must be produced in the brief moments during which the exploding star is dispersing its innards to space, mostly by capture of free neutrons.

3.5 Neutron stars and black holes

What happens to the compressed core of a supernova after its outer layers explode into space? The answer to this question was unknown until 1967. In that year *neutron stars* – the most interesting objects to be found in the sky for many years – were discovered.

When a large star collapses at the end of its life, just before it explodes as a supernova, the materials of the collapsing star pile up at the center, and produce enormous pressures, even more powerful than the inward pressure produced by the star's own weight. Under this crushing burden, a pure ball of neutrons forms at the center of the star, only 10 miles (16 kilometers) in radius, but with much of the star's original mass packed into it. The hypothetical ball of neutrons is a neutron star – a fantastically compressed, super-dense ball of matter, created when a massive star collapses at the end of its life.

This is a startling conclusion. The white dwarf – about 10 000 miles (16 000 kilometers) in diameter – was formerly thought to be the smallest, densest star in the Universe. How can a star be smaller than a white dwarf? The matter in such a star would be a billion times denser than the matter in a white dwarf. The density of matter in a neutron star is so great that a tablespoon of neutron star matter would weigh a billion tons (a trillion kilograms).

A dense shower of neutrinos accompanies the formation of the neutron star. Neutrinos, like neutrons, are electrically neutral; they carry no charge. But unlike the massive neutrons, neutrinos are very light. In fact, they may even be massless. These chargeless and massless (or nearly massless) particles interact very weakly with matter. A neutrino can pass through the entire body of Earth as easily as a ray of light passes through a pane of transparent glass.

Because neutrinos pass through material so easily, the shower of neutrinos emitted from the ball of neutrons at a supernova's center escapes through the outer layers of the exploding star without difficulty and spreads out into space. Thus, a supernova collapse and explosion should produce three important phenomena: a dense clump of neutrons – a neutron star – at the supernova's center; an outward-moving cloud of gas containing most of the innards of the original star; and an intense burst of neutrinos.

This theory was speculative for many years. Not since the dawn of modern astronomy had a supernova exploded close enough to our Solar System to provide a good check on these predictions. Then, in 1987, the long-awaited event occurred. A massive star, nearly 20 times more massive than the Sun, collapsed and exploded in a nearby dwarf galaxy called the Large Magellanic Cloud.

The Large Magellanic Cloud is a small satellite galaxy of the Milky Way, held captive by the force of gravity. It is only 160 000 light-years away – not quite in our Galaxy, but close enough to permit astronomers to observe it in detail (Figure 3.10).

On a summer night in 1987, a night assistant at Las Campanas Observatory in Chile looked up at the southern sky and saw a bright star he had never seen before. It was the supernova (Figure 3.11).

FIGURE 3.10 The Large Magellanic Cloud and Supernova 1987a. In this close-up of the Large Magellanic Cloud, the glowing cloud on the left is known as the Tarantula Nebula, and surrounds a cluster of hot, young stars. The bright "star" (lower right) is Supernova 1987a .

© Anglo-Australian Observatory/David Malin Images

FIGURE 3.11 Supernova 1987a after 10 years (Hubble Space Telescope image taken in 1997). The rings represent material expelled from the star prior to the supernova. At present, the expanding debris from the stellar explosion is colliding with this previously expelled material. The violent collision heats the ring material causing it to glow.

© Chandra X-ray center, Harvard University

Around the same time, bursts of neutrinos were recorded in two Earth-based instruments set up to trace their elusive paths. Both were located deep underground – one in a lead mine in Japan, and the other in a salt mine near Cleveland, Ohio. On the basis of these observations, astronomers calculated that 100 trillion neutrinos had passed through the body of every person on Earth.

Since the supernova had occurred in the Southern Hemisphere sky, and the neutrino detectors were in the Northern Hemisphere, all the neutrinos they detected must have passed through the body of Earth. The detection of the neutrinos provided striking confirmation of one of the key predictions of the supernova theory – the appearance of a burst of neutrinos as enormous pressures create the neutron star at the center of the supernova.

3.6 Pulsars

Another discovery made in the 1960s also points to the existence of a dense ball of neutrons at the center of an exploded star. The discovery came about by pure chance. Jocelyn Bell, then an astronomy student at Cambridge University, was assigned the task of investigating fluctuations in the strength of radio waves from distant galaxies. She found unexpectedly that certain places in the heavens were emitting short, rapid bursts of radio waves at regular intervals. Each burst lasted not more than one-hundredth of a second. The rapid succession of bursts seemed like a speeded-up celestial digital code.

The interval between successive bursts was extraordinarily constant. In fact, it did not change by more than one part in 10 million. A clock with this precision would gain or lose no more than a second a year.

No star or galaxy had ever been observed to emit signals as bizarre as these. At first, some astronomers thought that intelligent beings on other stars might be beaming a message to Earth, and referred to the Morse-code stars as LGMs, or "Little Green Men." But it soon became evident that the radio pulses had a physical origin. One clue was that the signals were spread over a broad band of frequencies. If an extraterrestrial society were trying to signal other solar systems, its interstellar transmitters would require enormous power in order for the signals to carry across the trillions of miles that separate every star from its neighbors. The only feasible way to do this would be to concentrate all available power at one frequency, as we do when we broadcast radio and TV programs. It would be wasteful, purposeless, and unintelligent to diffuse the power of the transmitter over a broad band of frequencies.

This cold reasoning dashed the hopes of romantics who believed for a short time that humans might be receiving their first message from outer space. LGMs disappeared from scientific conversation; *pulsars* took their place and scientists settled down to search for a natural explanation of the peculiar signals.

The next clue to the answer was the sharpness of the pulse. From the fact that each pulse lasted for one-hundredth of a second or less, astronomers concluded that a pulsar is an incredibly small object for a star, far smaller than a white dwarf. When an object emits a burst of radio waves, the waves from different parts of the object arrive at Earth at different times, blurring the sharpness of the original pulse. The smaller the object, the sharper the pulse. Following this line of reasoning, astronomers calculated that the objects were not more than 10 miles (16 kilometers) in radius.

Only one object was thought to have the mass of a star packed into a radius of about 10 miles (16 kilometers). That object is the neutron star. Could it be that the objects known as pulsars actually were neutron stars?

Motivated by this thought, in the 1960s astronomers searched assiduously for further evidence linking neutron stars and pulsars. They investigated with particular care the region at the center of the Crab Nebula, where the squeezed-down core of the supernova explosion of AD 1054 should have been located.

In 1968, a wave of excitement spread through the astronomical community when a pulsar was discovered at the center of the Crab Nebula precisely where the squeezed remnant of the star should be located. Suddenly, many items of evidence fitted together like the pieces of a jigsaw puzzle. Neutron stars are supposed to exist at the centers of supernovas; the Crab Nebula is the remains of a supernova; a pulsar has been discovered at the center of the Crab Nebula; and the neutron star and the pulsar are the only objects known that have the mass of a star packed into a 10-mile sphere. Clearly, neutron star and pulsar were two names for the same thing: a super-dense ball of matter, created when a massive star collapses at the end of its life.

What produces the sharp, regularly repeated bursts of radiation from which pulsars derive their name? The answer is believed to be that a pulsar, like the Sun and most other stars, is subject to violent storms that may last for years, spraying particles and radiation out into space. Each storm occurs in a localized area on the surface of the pulsar and sprays its radiation into space in a narrowly defined direction. When Earth lies in the path of one of these streams of radiation, our radio telescopes pick up the signals, which indicate to us the presence of the pulsar.

But if the spray of radiation is emitted steadily from the pulsar, why do we observe it as a succession of isolated sharp bursts? The reason is probably that pulsars, like most stars, spin on their axes. In fact, being smaller than normal stars, pulsars can spin very rapidly, as many as several times a second. As the pulsar spins, the stream of radiation from its surface sweeps through space like the light from a revolving lighthouse beacon. If Earth happens to lie in the path of the rotating beam, it will receive a sharp burst of radiation once in every turn of the pulsar.

This theory can be tested, because all spinning objects slow down steadily in the course of time as a result of friction. Thus, the interval of time between successive bursts of radiation from the pulsar must increase. In 1969, this prediction was confirmed by the discovery that the time between successive pulses from the Crab Nebula pulsar was getting longer, at the tiny but measurable rate of one-billionth of a second per day. The pulsar theory had passed its test.

3.7 Black holes

With the discovery of the neutron star, many astronomers thought that the story of the birth and death of stars was complete. But there is reason to believe that the neutron star or pulsar is not the ultimate state of compression of stellar matter. If the star is excessively massive, the collapse at the end of its life can squeeze the star's core down to a size even smaller than that of a neutron star. When the core is compressed to a radius of 2 miles (about 3 kilometers), Einstein's theory of relativity predicts the occurrence of an extraordinary phenomenon.

According to Einstein's theory, energy and mass are equivalent. The equivalence is expressed in the famous equation:

$$E = mc^2,$$

where E is the energy of the light ray, m is its mass and c is the velocity of light. Reference is frequently made to this equation in calculating the energy of a nuclear bomb. What has this to do with the collapsing star? If energy is equivalent to mass, a ray of light which possesses electromagnetic energy must also possess mass, just as if it were a particle of matter. Thus, a ray of light emitted from a star will be pulled back by the star's gravity, as a ball thrown up from the surface of Earth is pulled back by Earth's gravity. When the star is normal in size – about a million miles in diameter – the force of gravity on its surface is not strong enough to keep the light rays from escaping, and they leave the star, although with somewhat less energy.

But if the core of the collapsing star is squeezed into a very small volume, the force of gravity on its surface is very great. Suppose the core is squeezed down to a radius of a few miles. At that point, the mass is so compact that the force of gravity at the surface is billions of times stronger than the force of gravity at the surface of the Sun. The tug of that enormous force prevents the rays of light from leaving the surface of the star; they are pulled back and cannot escape to space. All the light within the star is now trapped by gravity. From this moment on, the star is invisible. It is now a *black hole in space.*

Inside the black hole, the contraction continues, piling up matter at the center in a tiny, incredibly dense lump. According to current knowledge in theoretical physics, this is the end of the star's life. The star's volume becomes smaller and smaller; from a globe with a 2-mile radius it shrinks to the size of a pinhead, then to the size of a microbe, and, still shrinking, passes into the realm of distances smaller than any ever probed. At all times, the star's mass of a thousand trillion trillion tons remains packed into the shrinking volume. But intuition tells us that such an object cannot exist. At some point the collapse must be halted. Yet, according to the laws of physics, no force, no matter how powerful, can stop the collapse. The implication is that the laws of physics must be modified at extremely short distances in a manner that prevents particles from coming infinitely close together.

Black holes have many other remarkable properties. For example, the force of gravity within a black hole not only prevents light from escaping; it also prevents all physical objects from getting out of the hole. This property of black holes is another prediction of Einstein's theory, which asserts that no object can travel faster than light. If the black hole's gravity is so powerful that light cannot break its grip and escape to space, material objects cannot escape either. Everything inside the black hole is trapped there forever.

Any ray of light or physical object that enters the black hole from the outside is also trapped; it can never emerge again.

The interior of the black hole is completely isolated from the outside world; it can take in objects and radiation, but cannot send anything back. It is almost as though the material inside the black hole no longer belongs to our Universe.

Since black holes capture any material they encounter, the size of a black hole will tend to increase with time. The black hole is, in a sense, insatiable. As more matter enters it, the strength of its gravity increases, and therefore its sphere of influence expands. This property does not imply that black holes act as gravitational vacuum cleaners, pulling in all the matter around them. A spaceship can pass by a black hole safely as long as it does not come very close. However, if a ship heads directly for a black hole, it will enter the black hole and vanish.

The black hole is one of the most bizarre objects that has ever been conceived by the scientific mind. Yet the theory of relativity, combined with calculations on the evolution of massive stars, assures us that black holes must exist; whenever a very massive star undergoes a supernova explosion, a black hole must be left behind. How can this prediction be tested?

A test would seem to be impossible, since the black hole by its nature cannot be seen. However, discoveries made with satellites provide tentative evidence that black holes actually do exist. The satellites are equipped to measure X-rays coming to Earth from space. These X-rays cannot be observed on the ground because Earth's atmosphere screens them out. One very powerful source of X-rays discovered by the satellites seems to be coming from a bright, conspicuous star. However, a very careful observation shows that the X-rays are not coming from the star itself, but from an invisible object quite close to it.

The invisible object is thought by astronomers to be a black hole circling around a normal star. According to their calculations, the powerful gravitational pull of the black hole tears streamers of gaseous matter away from the neighboring star. The atoms of gas spiral in toward the boundary of the black hole, colliding violently on the way, and the collisions produce an intense stream of X-rays (Figure 3.12). Astronomers have tried to invent other theories to explain the mysterious X-rays; however, nothing fits all the known facts as well as the black hole.

FIGURE 3.1 2 Observation of a black hole. The powerful gravitational pull of the black hole draws gas from the outer envelope of a companion giant star. The in-falling gases, heated to high temperatures, emit streams of X-rays.

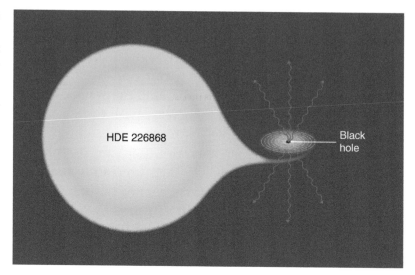

HDE 226868

Black hole

3.8 Quasars and giant black holes

Many years ago, astronomers noticed the existence of objects that seemed to be stars – at least, they were starlike points of light – but unlike ordinary stars, these objects were rather fuzzy at their edges. They were also extremely distant – far too distant to be ordinary stars. When the apparent brightness of these strange objects was corrected for their great distance from our Solar System, their true brightness turned out to be very great. All were many billions of times brighter than an ordinary star.

No single star could be that bright. These objects had to be galaxies – clusters of billions of stars. But why did they have the appearance of starlike points of light? Why did they not look like galaxies?

During the course of time, astronomers came to realize, through a succession of discoveries, that the strange objects were indeed whole galaxies of stars, and not single stars. However, these galaxies had extraordinarily bright centers. Because the galaxies were so far away, the ordinary outlying part of each galaxy was very faint, and did not show up on the astronomical photographs. But the region at the center of the galaxy showed up clearly as a small spot of light because this region was so extraordinarily bright. This was the explanation for the fact that the objects look like somewhat fuzzy, but still starlike, points of light. The unusual galaxies were dubbed quasi-stellar objects, or *quasars* for short.

The extraordinary brightness of quasars was only one of their unusual properties. Even stranger was the fact that their enormous outpouring of energy seemed to be coming from a remarkably small region in space – smaller, in fact, than our Solar System.

That was the real puzzle. How can an object as small as a solar system produce the energy of hundreds of billions of stars? If quasars produce the energy of hundreds of billions of stars, they must contain hundreds of billions of stars. But quasars occupy far too small a volume of space to contain that many stars.

Astronomers wondered if a force as yet unknown to science was generating the energy of the quasars – a force even stronger than the nuclear force that generates the energy radiated by an ordinary star. That would be a momentous finding, for up to now the nuclear force has been the most powerful force known to man. The mystery deepened.

But then a clue to the nature of the strange quasars appeared. When astronomers modified the instruments in their telescopes so that they could detect very dim objects, they saw the faint image of a galaxy around several of the quasars. At first, astronomers had not noticed the galaxy surrounding each of these quasars, because each galaxy was so much fainter than the brilliant quasar at its center. If the astronomer set the time exposure in the telescope short enough to avoid overexposing the image of the bright quasar, the galaxy's image failed to appear in the photograph.

As soon as it was realized that quasars were located in the centers of galaxies, several astronomers thought of a solution to the mystery. What is extremely compact yet can generate an enormous amount of energy? Only one object fills the bill. It seems almost too bizarre to be true, but – is it possible that an enormous black hole lurks in the center of each of these galaxies? Are quasars really giant black holes?

If quasars are giant black holes, the source of the quasar's enormous energy output becomes clear. The black hole sits in the center of its galaxy, surrounded by many stars. The stars circle around the black hole under the pull of its gravity. Gradually they spiral in toward it. As each star draws close to the black hole, its gaseous body is torn apart by the black hole's powerful gravitational force. The atoms of gaseous matter within the disintegrating star, picking up speed under the attraction of the black hole, move faster and faster. As the atoms approach the boundary of the black hole, they collide with one another. The collisions heat the gas, and the hot gas radiates energy into space. This energy is what we see when we observe a quasar.

The calculations show that if the black hole in the center of the galaxy is an ordinary-sized black hole – the kind that results from the explosion that marks the death of a massive star – it will not produce enough energy to explain the brilliance of the quasars. But if the black hole is giant-sized – say, a billion times more massive than an "ordinary" black hole – it tears apart nearby stars with such vigor that it radiates huge amounts of energy into space – sufficient to account for the energy coming from even the most dazzling quasars.

So, the presence of a giant black hole lurking at the center of a galaxy explains the two remarkable properties of quasars. First, it accounts for the prodigious amount of energy coming from a quasar. Second, since black holes are exceedingly compact objects, it accounts for the fact that the energy comes from a small region in space.

As soon as astronomers conceived the idea that giant black holes might exist in the centers of galaxies, they found it easy to understand, with hindsight, why the center of a galaxy was exactly where a giant black hole ought to be.

In the center of a galaxy, stars are much more closely packed together than they are anywhere else in the galaxy, because the galaxy's gravitational pull tends to draw them toward the center, and they pile up there. The density of the stars near the center of the galaxy can be millions of times greater than in its outer regions. Since the stars are packed together so closely in the center of the galaxy, they collide there frequently. In the outer part of a galaxy, collisions between stars are rare; perhaps one occurs every billion years. But near the center, a collision between stars may occur every hour.

When two stars collide, they tend to fuse into one larger star in place of the two separate, smaller ones. Collisions being frequent in the center of the galaxy, this newly enlarged, double-sized star is likely to collide again in a short time with a third star, producing a still larger star as a result. Several collisions in a row lead to a very large and massive star – the kind that ends its life in an explosion, leaving behind a black hole.

That means the center of a galaxy should contain a considerable number of black holes. These black holes are not yet giant-sized, they are the "ordinary" black holes formed when a massive star explodes and dies. But Einstein's theory of relativity explains why a large number of "ordinary" black holes in the center of a galaxy are likely to amalgamate into a single giant-size black hole.

The theory predicts that the diameter of a black hole is proportional to the amount of matter inside it. Thus, every time a black hole encounters another object and swallows it, the black hole grows bigger. Being bigger, the black hole is now even more likely to collide with and swallow other objects. A runaway process starts in which the bigger the black hole is, the more likely it is to swallow other objects, and the more objects it swallows, the bigger it gets.

The runaway process continues until the largest black holes have swallowed up all the smaller ones. Finally the remaining black holes collide with and swallow one another. At the end, one giant black hole sits in the center of the galaxy.

After the giant black hole has formed in this way, the stars of the galaxy continue to circle around it, gradually drawing nearer. Once in a while, a star comes too close. Pulled in by the giant black hole's gravity, it is torn apart and consumed. The ravished star emits a great burst of energy in the final moments of its existence. These bursts of energy, coming one after the other as the stars close to the black hole are consumed, fuel the quasar's extraordinary energy output.

The story of quasars and giant black holes is nearly complete. A quasar is a galaxy with a giant black hole at its center. The quasar's dazzling radiation is created by stars that feed one by one into the massive black hole. Each time the giant black hole tears a star apart, we see the quasar flare up as if

FIGURE 3.13 Quasar 3C273, at upper left, is 3 billion light-years from Earth and is roughly 50 times brighter than the average galaxy. The energy emitted from this quasar comes from gas that falls inward toward the massive black hole at its center. The image shows a stream of material 250 000 light-years long that has been violently ejected from the core of the quasar at close to the speed of light.
    NASA/CXC/SAO/H. Marshall *et al.*

another log has been thrown on the fire. Some quasars are seen to emit streams of matter that travel at velocities close to the speed of light (Figure 3.13).

At first, the fire blazes brightly because the giant black hole has an ample supply of stars available for feeding. In other words, the brilliance of the quasar is very great in this early period.

But after a time, many stars in the inner part of the galaxy are gone because they have been torn apart and consumed by the giant black hole. After a relatively short interval of time, perhaps a few hundred million years, very few stars are left. With its source of energy gone, the quasar fades into darkness. Where the quasar once blazed, a galaxy of ordinary appearance remains – but with a quiescent black hole slumbering at its center.

Earlier in the history of the Universe, quasars – that is, galaxies with active, star-eating black holes at their centers – may have been fairly common. In fact, many galaxies may have been quasars then, with giant black holes at their centers, consuming nearby stars. But in these galaxies the black holes have long since eaten all the stars within their reach, and are no longer active. The quasars have faded into ordinary galaxies.

In that case, why do we see any quasars at all? Why hasn't every quasar run out of fuel by now, and faded away? The answer is connected with the fact that most of the quasars we see are very far away. Because they are so distant, it takes a long time for the light from these quasars to reach us. Thus, we see these quasars not as they are today but as they were in the distant past, when the light from them had just started on its journey to Earth. In that early epoch, the giant black holes at the centers of those galaxies had not yet consumed all the stars around them, and the quasars were still burning brightly.

Today those black holes are slumbering, and the quasars are dark. If we could see one of them as it is now, we would only see a galaxy – but we cannot see it as it is now, because it is so far away. We can only see it as it was long ago.

If this explanation is correct, many galaxies may have giant black holes lurking in their centers. If a galaxy is close to us, we see it as it is today; the giant black hole at its center is no longer active, and it looks like an ordinary galaxy. But if the galaxy is distant enough, we see it as it was in the past, when its giant black hole was active – that is, when it was a quasar.

With the evidence for the existence of black holes, the final pages in the life story of the stars appear to have been written. But the story has an epilogue. When a supernova explosion occurs and the outer layers of the star are sprayed out into space, they mingle with fresh hydrogen to form a gaseous mixture containing all 92 elements. Later in the history of the Galaxy, other stars are formed out of clouds of hydrogen which have been enriched by the products of these explosions. The Sun is one of these stars; it contains the debris of many supernova explosions dating back to the earliest years of the Galaxy. The planets, including Earth, are composed almost entirely of this debris. We owe our corporeal existence to events that took place billions of years ago, in stars that lived and died long before the Solar System came into being.

3.9 Summary

By the end of this chapter you should be familiar with the following concepts and topics:

The life cycle of a star
Formation of a star
Stellar evolution
Life of a star
   Formation of the elements now contained in our bodies
   Predominance of lighter elements in the Universe
Explosion of massive stars in supernova events
   Creation of heavy elements (silver and gold)
   Neutron stars
   Pulsars
   Black holes
   Quasars

Questions

3.1. As a gas cloud contracts by the force of gravity to form a star, its temperature rises. Why? What is the critical temperature at which nuclear reactions begin, and what happens at that point?

3.2. Why does a large star live for only millions of years while a small star may live for trillions of years?

3.3. Explain why hydrogen and helium are the most abundant elements in the Universe.

3.4. Why is iron surprisingly abundant in the Cosmos?

3.5. Why are the elements heavier than iron very scarce?

PART II

The Solar System

CHAPTER 4

Formation of the conditions for life

Planets and moons

The next part of our story on the long road to life is the formation of our Sun and Solar System. By four and a half billion years ago, innumerable supernova explosions had caused the concentration of heavier elements in the Universe to build up to about one percent of all matter, and conditions were set. In a spiral arm of the Milky Way Galaxy, a cloud of gaseous matter began to condense. In the hot center of this cloud, our Sun was formed, and in the cooler and less dense outer regions, our planets were born. The Solar System, and conditions allowing life to evolve, were starting to take shape. Astronomers believe that whenever a star is formed, planets may form around it. If this is true, there may be millions of planets in our Universe, many with potential for the existence of life.

In this chapter we will learn how the planets in our own Solar System formed 4–6 billion years ago from a rotating cloud of gas and dust. As the cloud heated up, small terrestrial planets accreted from rocky material in the warm inner regions, and in the cold outer regions the giant gaseous planets formed. After 10–20 million years, planetary formation was complete. We will take a tour of the incredible results, and learn about the diverse features of the planets, their satellites, and the asteroids, meteoroids, and comets that together form our Solar System.

4.1 The origin of the Sun and planets

How did the planets form? The modern theory of the birth of stars suggests that whenever a star is formed, planets may form around it. If the theory is correct, millions upon millions of planets exist in the observable Universe.

In this picture, when our Solar System was young, it consisted of a cloud of gas, mainly hydrogen and helium, in which small grains of solid matter were embedded. The grains included bits of iron, rock, and icy substances formed from elements made earlier in other stars and dispersed to space in supernova explosions.

As described earlier, such clouds, condensing out of the mists of space under the pull of gravity, are the first stage in the birth of every star.[1] With the passage of time, the cloud continued to contract under its own gravitational pull; as it contracted, the cloud began to rotate faster and faster. This happens because such a cloud, called a *nebula*, would be likely to have some intrinsic circular motion. As the cloud contracts, it rotates faster in much the same way as a spinning ice skater spins faster as she pulls her arms closer to her body.

The rapid rotation caused the cloud to bulge out and flatten along the plane of its rotation. Eventually, the cloud developed into a large disk. The cloud also grew hotter, until finally the temperature in the center reached the critical level of 10 million degrees Celsius needed for the ignition of

---

[1] The initial increase in density that leads to the collapse of a cloud may be triggered by the shock wave created by a nearby supernova. The presence of tiny mineral grains that are older than the Solar System in some primitive meteorites supports the idea that a supernova explosion initiated the formation of the Solar System.

Previous page: The Solar System (not to scale) showing the Sun, Inner Planets, Asteroid Belt, Outer Planets, Pluto (originally classified as a planet), and a comet.
  NASA.

thermonuclear reactions. This point in the contraction marked the birth of the Sun. In our Solar System, this event occurred about 4.6 billion years ago.

As the Sun was forming out of the hot center of the cloud, smaller knots of condensed matter appeared in the disk-shaped cloud's cooler outer regions. According to this theory, the smaller knots of matter became the giant planets (Figure 4.1). Recent studies suggest that the cores of the giant planets formed rapidly by the collection of rock and ice particles in the outer part of the nebula, which then drew in the surrounding gases by the force of gravity.

This theory provides a plausible explanation for the birth of the so-called *giant planets* in the Solar System – Jupiter, Saturn, Uranus, and Neptune – which are composed mainly of hydrogen and helium, the same gases that make up the bulk of the Sun, with lesser amounts of water, methane, and ammonia.

However, this explanation of planetary origins, which seems to work well for the giant planets, does not work so well for Earth and its sister planets – Mercury, Venus, and Mars. These bodies, called *earthlike* or *terrestrial planets*, are very different from the giant planets. They are largely composed of rock and iron, with very little hydrogen and helium. They are also much smaller than the giant planets; Earth, for example, has a mass of only 1/300 that of Jupiter.

Then how did the terrestrial planets form? To find the answer, consider what must have happened to the solar nebula with the passage of time. The nebula began to cool, and when it had cooled sufficiently, atoms would begin to stick together to form tiny grains of solid matter. The first grains of matter to appear in abundance were bits of iron. Next, silicate minerals, bits of rocklike material, composed largely of the elements silicon and oxygen condensed. As the cooling continued, more bits of rocky material appeared. Finally, water appeared in the form of ice crystals.

(a)

(b)

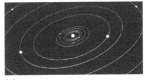
(c)

FIGURE 4.1 Formation of the planets. Terrestrial planets accrete from rocky material in the warm inner regions of the solar nebula. Meanwhile, the huge, gaseous Jovian planets form in the cold outer regions. (a) The solar nebula in its initial stages. (b) The early Solar System. (c) Planetary formation is nearly complete after only about 10 to 20 million years.

From *Universe*, 5/e by William J. Kaufmann III and Roger A. Freedman. © 1985, 1988, 1991, 1994, 1999 by W. H. Freeman and Company. Used with permission.

4.1.1 The Sun's flare-up

Now we come to a key point in the theory. Astronomers have observed that stars like the Sun flare up frequently when they are very young, sending sporadic bursts of particles and radiation into nearby space. They believe that when the Sun was young, probably less than a million years old, its surface erupted in violent outbursts, evaporating and expelling gases and small bits of matter from the inner part of the parent cloud. But the larger pieces of rock and iron were too heavy to be blasted away by the star's outbursts. They remained after the gases had been driven off.

Gases must have been removed at least as far out as the orbits of the asteroids, rocky bodies which lie between Mars and Jupiter. If hydrogen and helium were still present in their full abundance in the asteroid belt when the asteroids formed, then a gas giant like Jupiter would probably have formed there. That is, a planet whose composition was mainly hydrogen and helium would have appeared. The fact that no planet of this kind is found in the asteroid belt indicates that the gases of the solar nebula were removed at least as far out as that distance from the Sun.

4.1.2 Accretion: Planetesimals into planets

Now Earth and its sister planets in the inner part of the Solar System began to form out of those fragments of iron and rock. Collisions occurred now and then between neighboring chunks of rock and iron as they orbited the Sun. When the collisions were gentle enough, the chunks might stick together. In this way small grains of rock and iron rapidly grew into larger ones, making kilometer-sized bodies called *planetesimals*. Eventually, some planetesimals became large enough to exert a gravitational attraction on their neighbors. Once that happened, these larger planetesimals quickly swept up all the smaller fragments of matter in the space around them and developed into full-sized, earthlike planets. Finally, nearly all the material of the primordial accretion disk was gathered into planets Mercury, Venus, Earth, and Mars. This is the situation in the Solar System as it exists today.

4.2 Other solar systems

Observations of planetary systems in the act of formation show characteristics that agree with the theoretical results.

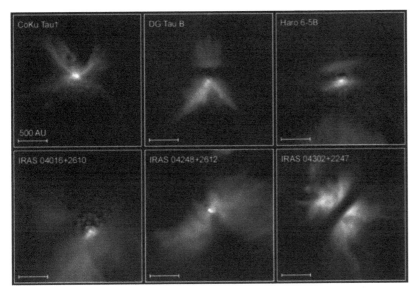

FIGURE 4.2 Edge-on views of accretion disks around young stars (as seen by the infrared camera of the Hubble Space Telescope). The central objects, now newly formed stars, are shrouded in dust. The wisps of gas surrounding the young stars are glowing from reflected starlight and gas is being ejected along the axis of rotation of the disk.

D. Padgett (IPAC/Caltech), W. Brandner (IPAC), K. Stapelfeldt (JPL) and NASA

Astronomers have observed accretion disks around a number of young stars (Figure 4.2). These newly formed stars are typically imbedded in a dense dust cloud, and gases are ejected at a high velocity along the axis of rotation of the accretion disk.

Since 1995, several groups of astronomers have reported the detection of planets circling other stars. The discovery of these other solar systems rests on the fact that just as a star's gravitational pull keeps a planet in orbit around it, the planet will also have a much smaller, but perceptible pull on the star. Thus, a planet circling its star in a somewhat elliptical orbit should cause the orbit of the star to wobble. In reality, both planet and star move in elliptical orbits around a common point.

As seen from Earth, the star would appear to move back and forth as it orbits the center of mass of the system. The wobble will cause the star to be moving back and forth in space, at times moving toward us and at times away from us. This movement is too small to be seen, but can be detected. As discussed previously, light from a source moving toward an observer should show a blue shift and light moving away should show a red shift. Light from the wobbling star would be alternately blue-shifted and red-shifted, with a characteristic period and amplitude of oscillation that reveal the mass

FIGURE 4.3 The arrangements of planets in orbit around some nearby stars. For comparison, the arrangement of the inner planets in our own Solar System is shown at the top of the figure. The scale at the bottom of the figure shows the average distance from each planet to its star in astronomical units (1 AU is the Earth–Sun distance). The planets themselves are not shown to the same scale as their stars, and the relative sizes of the extrasolar planets are estimates only. Extrasolar planetary systems may in many cases have additional undiscovered planets. The mass of each planet is given as a multiple of Jupiter's mass (M_{Jup}).

Courtesy G. Marcy

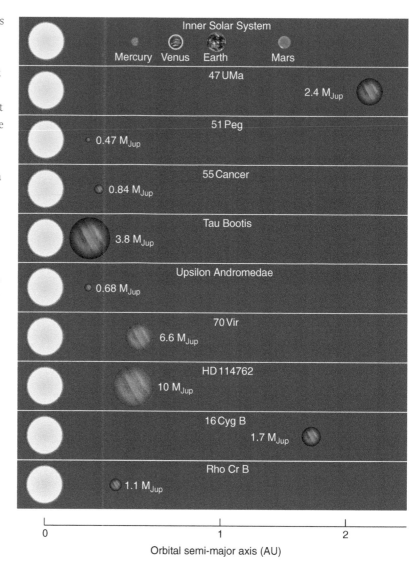

and orbital distance of the planet or planets causing the perturbations. The method requires sensitive spectroscopes that can measure very small Doppler shifts, and with the development of new instruments in the mid 1990s, the first planets orbiting other stars were discovered. To date, more than 135 stars have been found to have planets orbiting around them (Figure 4.3).

Thus far, many of the planetary systems that have been detected are quite different from our own, with giant planets orbiting very close to their parent stars. It appears that in some planetary systems, giant planets form while there is still a great deal of gas in the surrounding nebula. Friction between the planet and the gaseous nebula slows the planet's orbit and it spirals in toward the central star.

Some astronomers have taken this as evidence that solar systems like ours, with small terrestrial planets relatively close to the central star, are rare. Others contend that the first planets discovered would naturally be those that are larger and close to the stars as they are the easiest to detect. The discovery of a planet around the star 16 Cygnus B with less than twice the mass of Jupiter at almost twice the Earth–Sun distance provides evidence of systems more like our own. As the spectroscopes become somewhat more sensitive, we should begin to see solar systems more like our own.

4.3 A tour of the terrestrial planets

The planets of the Solar System divide into two groups differing greatly in their size, mass, and composition (Figure 4.4). Mercury, Venus, and Mars resemble Earth in being composed almost entirely of rocky materials and iron, and are known, together with Earth, as the terrestrial planets. The Moon is only slightly smaller than Mercury, and is sometimes included with the terrestrial planets, being composed of similar materials. (The Moon, Earth, Mars, and Venus will be discussed in detail in subsequent chapters.)

4.3.1 Mercury

Mercury has a diameter one-third the size of the Earth's and has only one-twentieth Earth's mass. Orbiting the Sun at a distance of about one-third of Earth's distance, it completes one orbit in 88 days. Its period of rotation is approximately 69 days. A close view of Mercury obtained in 1975 from the Mariner 10 spacecraft (Figure 4.5) revealed a heavily cratered moonlike surface. Mercury resembles the Moon in lacking an atmosphere, liquid water, and certainly life. Daytime temperatures at the equator are above 400 °C (about 750 °F) – hot enough to melt lead, whereas nighttime temperatures drop precipitously.

FIGURE 4.4 The Sun and planets, arranged in order of distance from the center of the Solar System, are shown here in proportion to their actual sizes. The Sun is 860 000 miles (1.4 million kilometers) in diameter, or roughly 10 times the size of the largest planet, Jupiter.

Lunar and Planetary Laboratory

FIGURE 4.5 The heavily cratered moonlike surface of Mercury. This mosaic of Mercury was composed from dozens of images recorded by the Mariner 10 spacecraft. The blank swatch running down from Mercury's north pole is a region that was not imaged by Mariner 10, which imaged about 50% of the surface of the planet.

USGS

Recent studies suggest that water in the form of ice may exist in some deeply shadowed craters near Mercury's poles (Figure 4.6).

4.3.2 Venus

Venus is Earth's nearest neighbor, and also has nearly the same size and mass as Earth (Figure 4.7). The planet circles

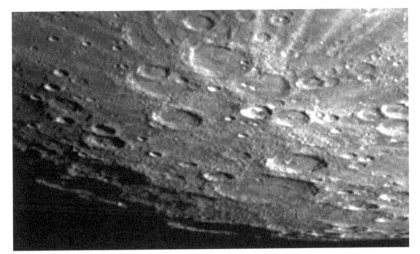

FIGURE 4.6 Shadowed craters of Mercury as photographed by the Mariner 10 spacecraft. Earth-based radar observations have detected unusually reflective patches within some craters in the polar regions of Mercury. Some astronomers believe that this represents water ice preserved in areas permanently shadowed from sunlight.

NASA/JPL/Northwestern University

FIGURE 4.7 An ultraviolet image of Venus showing cloud patterns, obtained by the Pioneer Venus Orbiter spacecraft in 1979.

NASA/NSSDC

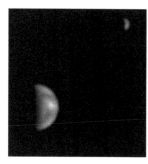

FIGURE 4.8 Earth and its Moon imaged by the camera onboard the Mars Global Surveyor spacecraft. Features visible on Earth include the Pacific Ocean, clouds, much of South America, and part of North America. Earth's Moon is visible on the upper right, with the crater Tycho brightening the lower part.

David Crisp and the WFPC2 Science Team (Jet Propulsion Laboratory/ California Institute of Technology) and NASA

the Sun at about two-thirds the distance of Earth's orbit, completing one circuit in 225 Earth days.

The planet's slow rotation on its axis (one Venus day is equivalent to 243 Earth days) is opposite to the direction of its movement around the Sun. That is, looking down on the north pole of Venus from "above", the planet will be seen to revolve about the Sun in a counterclockwise direction – as is the case for all planets in this Solar System – while rotating on its axis in a clockwise direction. This type of motion is called *retrograde rotation*. Venus is completely covered in clouds, and has a dense carbon-dioxide rich atmosphere and a surface temperature of about 500 °C (about 900 °F).

4.3.3 The Earth–Moon system

Earth and its Moon orbit the Sun at a distance of 93 million miles (150 million kilometers), the definition of one Astronomical Unit (AU). Earth's diameter is 8000 miles (12 756 kilometers) and the smaller Moon is 2160 miles (3476 kilometers) in diameter. The Moon lies an average of 238 900 miles (384 400 kilometers) from Earth (Figure 4.8).

Major characteristics of Earth are its geological activity, an abundance of liquid water, and an atmosphere. The Moon has no liquid water, no atmosphere, and no evidence of recent geological activity. The heavily cratered surface of the Moon records the history of bombardment of the planets by asteroids and comets. On Earth, this record has been largely destroyed by erosion and deposition by water and wind. In Chapters 5, 8, and 9, we will discuss the geological evolution of Earth and its Moon in detail.

4.3.4 Mars

Mars orbits the Sun at a distance of 142 million miles (227 million kilometers), one and one-half times as distant as Earth. The length of the planet's year is 687 days or 1.9 Earth years. It rotates on its axis in 24 hours and 37 minutes. Mars has a thin atmosphere of carbon dioxide and the planet is cold and dry.

The most conspicuous features in Earth-based photographs are its polar caps, resembling the cover of ice and snow at the poles of Earth. Spacecraft measurements have indicated that the caps are composed of a thin layer of frozen carbon dioxide (dry ice) covering a less extensive but thicker cap of water ice (Figure 4.9).

Mars has two small moons named Phobos and Deimos (Figure 4.10). Phobos circles the planet in a close orbit that carries this moon around Mars in $7\frac{1}{2}$ hours – shorter than a Martian day. These miniature moons are unusually dark in color and may be captured carbon-rich asteroids.

FIGURE 4.9 The change of seasons on Mars. This Hubble Space Telescope image was made in March 1997, on the last day of spring in the Martian northern hemisphere. Like Earth, Mars has seasons that are caused by the tilt of the planet's rotation axis.
Phil James (University Toledo), Todd Clancy (Space Science Institute, Boulder, CO), Steve Lee (University Colorado), and NASA

4.4 A tour of the outer planets

Five planets lie outside the orbit of Mars. These are the giant planets – Jupiter, Saturn, Uranus, and Neptune – and the small planet Pluto.

The giant planets are approximately 3 to 10 times larger in diameter than Earth and 10 to 100 times more massive, but considerably lower in density. In general, their density is about the same as that of water: Saturn, in fact, is less dense than water; it would float in the bathtub if you could get it in. The giant planets are less dense than Earth and its neighbors because they contain large amounts of the lightest elements, hydrogen and helium.

4.4.1 Jupiter

Jupiter is the largest of the giant planets with a mass more than 300 times that of Earth (Figure 4.11). Its average distance from the Sun is 480 million miles (770 million kilometers), about five times the Earth–Sun distance. Jupiter completes a revolution around the Sun in approximately 12 Earth years. The length of its day is 9 hours and 50 minutes. The rate of spin of Jupiter is very rapid for a planet of its size, and this leads to a pronounced bulge of material at the equator. The other giant planets also have large equatorial bulges.

Theoretical estimates of the internal structure of Jupiter indicate that the temperature at the center may be as high as 50 000 °C (about 90 000 °F). The high temperature is a result of the compression of this very massive planet under the force

(a)

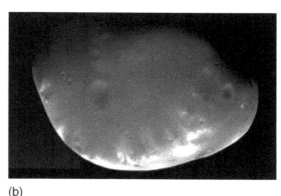

(b)

FIGURE 4.10 The moons of Mars.

(Left) Phobos, the larger of Mars' two moons, is potato-shaped with a maximum diameter of about 17 miles (27 kilometers).

NASA/JPL/Malin Space Science Systems

(Right) Deimos seems to be less cratered than Phobos, with a maximum diameter of only 10 miles (16 kilometers).

Viking Project, JPL, NASA

of its own gravity. If Jupiter were roughly 50 times more massive, its internal temperature would have risen to a level sufficient to ignite nuclear reactions converting it into a small star.

It might be expected that the interior of Jupiter would be in a gaseous state resembling the interior of a star, in view of the high temperature at its center. However, calculations on the properties of hydrogen, when subjected to enormous pressures and a temperature of 50 000 °C, indicate that the extreme pressure forces the atoms into the state of a liquid metal, like lithium or sodium at high temperatures. Jupiter is also believed to have a core of rocky material (Figure 4.12).

A thick envelope of highly compressed, gaseous hydrogen and helium overlies the hydrogen core, extending upward with gradually diminishing density. The gaseous envelope is topped by two thick decks of clouds. The lower deck consists of water droplets and ice crystals. The upper layer of clouds consists of crystals of frozen ammonia (NH_3) compounds at a temperature of -150 °C (-238 °F). These clouds of frozen ammonia present the visible face of the planet to the observer.

Photographs taken from spacecraft show turbulent cloud formations that appear to be violent storms extending over

FIGURE 4.11 Jupiter. This photograph of Jupiter was taken by cameras on the Voyager 1 spacecraft in 1979 at a distance of 15 million miles (24 million kilometers). The bands circling the planet are belts of moving clouds made of crystals of frozen ammonia (NH_3). The temperature at cloud-top level is about −150 °C (−238 °F). A zone of water clouds and warmer temperatures lies beneath the ammonia cloud cover. The great Red Spot (lower center) and several smaller white spots are mammoth storm centers.

NASA Jet Propulsion Laboratory (NASA-JPL).

thousands of kilometers (Figure 4.13). The photographs also reveal details of the great Red Spot, about 20 000 miles (32 000 kilometers) long and 6000 miles (9600 kilometers) wide. The Red Spot is probably a giant storm center. However, the Red Spot has persisted for at least 400 years, whereas hurricanes and storms on our planet play themselves out and disappear after a few weeks at most as their energy is dissipated into turbulence in the surrounding air.

Conditions beneath the clouds of Jupiter are concealed from our view, but below the level of the cloud tops the temperature must rise. Jupiter does not have a well-defined surface but, instead, grows steadily denser with increasing depth.

4.4.2 The Galilean satellites

Jupiter has sixteen main satellites – four large moons and twelve smaller bodies that are most likely captured asteroids. Jupiter is also surrounded by many very small irregular

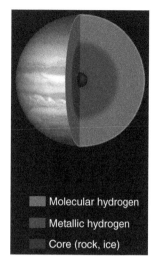

satellites, and by a faint ring system composed of very fine particles of rock that may originate from meteorite impacts on the small inner moons of the planet.

The four large moons, in order of increasing distance from Jupiter, are Io, Europa, Ganymede, and Callisto (Figure 4.14). These four moons, which were discovered by Galileo in 1610 when he first turned his telescope on Jupiter, are known as the *Galilean satellites*. Their relatively rapid motions around their planet suggested to Galileo that Jupiter was a solar system in miniature. Galileo concluded that Earth was not the center of all motion. His conclusion was a powerful argument for the view of Copernicus that Earth was not the center of the Universe, but moved instead around the Sun.

FIGURE 4.12 The internal structure of Jupiter. Jupiter probably has a rocky core. This core is surrounded by a thick layer of liquid metallic hydrogen and a relatively shallow outer layer composed primarily of ordinary hydrogen.

NASA

Io

Io is the innermost of the Galilean satellites. The first spacecraft observations of the surface of Io in 1979 revealed one of the greatest surprises in the history of planetary exploration: this satellite has active volcanoes (Figure 4.15). Several volcanoes were observed on Io by the Voyager spacecraft, and by the more recent Galileo spacecraft. There is evidence that

FIGURE 4.13 The Red Spot. This detailed view of the Red Spot (upper right) taken at a distance of 3 million miles from Jupiter, reveals patterns of violent turbulence in the atmosphere of the giant planet. The Red Spot, 20 000 miles (32 000 kilometers) in diameter, is believed to be a gigantic storm.

NASA

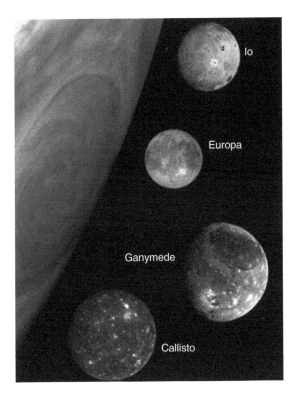

FIGURE 4.14 This montage of Galileo spacecraft images shows Jupiter and its four largest moons – Io, Europa, Ganymede, and Callisto. Io and Europa are composed largely of rocky materials like our Moon. Io has numerous active volcanoes, while Europa's icy surface may cover an ocean of liquid water. Ganymede has had a geologically active history. Callisto, by contrast, seems to be geologically dead.
NASA

volcanic eruptions take place continuously at one point or another on Io's surface (Figure 4.16).

The volcanoes on Io were a surprise because Io, while fairly large for a moon, is quite small in comparison to volcanically active bodies such as Earth or Venus. Volcanoes on a planet like Earth are believed to be the result of heat released in the interior of a planet by the decay of uranium and other radioactive substances. This heat raises the temperature of the rocks in the interior to the melting point. However, if the planet is small, the heat produced in its interior rises to the surface and is lost to space, and the planet becomes volcanically inactive in a relatively short time. Earth's Moon became volcanically inactive about 3 billion years ago, probably as a consequence of its small size and rapid rate of loss of internal heat. Io, being smaller than the Moon, should have lost its internal heat at a faster rate and become volcanically inactive even earlier. Why is Io still volcanic? What is the source of its heat?

The answer is connected with the pull of Jupiter's gravity. The gravitational attraction of Jupiter, tugging at the near side

FIGURE 4.15 Io as imaged by the Galileo spacecraft showing many volcanoes and volcanic calderas including the large volcano Pele (center, right). Pele is surrounded by a large red ring that represents deposits of sulfur from a previous eruption. The lava flows that cover the surface of Io are composed largely of various forms of molten sulfur.
    NASA/JPL-Caltech

of Io harder than the far side, raises tides on the surface of the satellite similar to the tides raised on Earth by the Sun and the Moon. The height of the tides depends on Io's distance to Jupiter, which varies during its orbit around the planet. Since the pull of the other Galilean satellites causes this distance to vary continually, the heights of the tides also vary. These continually changing tides bend and flex the body of the satellite, causing internal friction and heating. The frictional heat accumulated in this flexing motion, repeated in orbit after orbit, is sufficient to heat the rocks in the interior of Io to the melting point, leading to surface volcanism.

Europa
Europa is slightly smaller than Earth's Moon. It is an extra-ordinarily smooth body as a result of a surface coating of ice

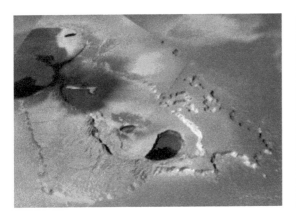 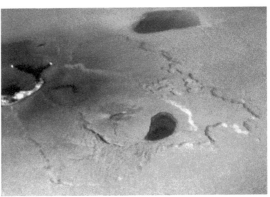

FIGURE 4.16 Tvashtar
Catena, a region on Io
marked by active volcanism.
The panel on the right shows
an active lava flow visible in
the infrared part of the
spectrum.
NASA/JPL-Caltech

(Figure 4.17). The icy surface shows myriad reddish brown markings that we now know to be cracks filled with material that came up from below. However, ice and water cannot make up a large fraction of its composition, because the density of Europa is 3 grams per cubic centimeter, which is approximately the same as the density of rock. A comparison with the Moon's density of 3.3 suggests that Europa is made of 90% rock and iron, and about 10% water.

The tidal forces produced by Jupiter must heat the interior of Europa in the same manner in which they heat Io, although not to as great a degree. Consequently, the water content of Europa probably exists in liquid form. It may form a deep global sea, capped by a relatively thin crust of ice (Figure 4.18). Recently, there has been speculation that Europa's ice-covered oceans may harbor life; perhaps primitive forms living around active submarine volcanic vents.

Ganymede and Callisto

Ganymede and Callisto are about as large as the planet Mercury, and are the second and third largest moons in the Solar System. Studies indicate that they consist of roughly equal parts of rock and water, the latter probably in the form of ice. Ganymede and Callisto are too far away from Jupiter to be heated appreciably by its tidal forces, but they contain enough rocky materials – and probably, therefore, enough radioactive elements – so that radioactive heating could have been appreciable at an earlier time. However, because of the relatively small size of these satellites compared to a planetary body like Earth, much of this heat should have

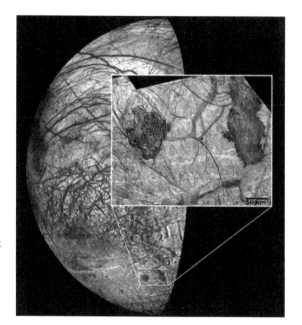

FIGURE 4.17 Europa as imaged by the Galileo spacecraft shows the ice-covered moon marked by numerous cracks filled with orange and brown materials that welled up from beneath the ice.

NASA/JPL-Caltech

FIGURE 4.18 Close-up of Europa taken by the Galileo spacecraft showing an area of broken and displaced icy crust. The disruption of the ice was probably caused by movement of water in the ocean below.

NASA/JPL-Caltech

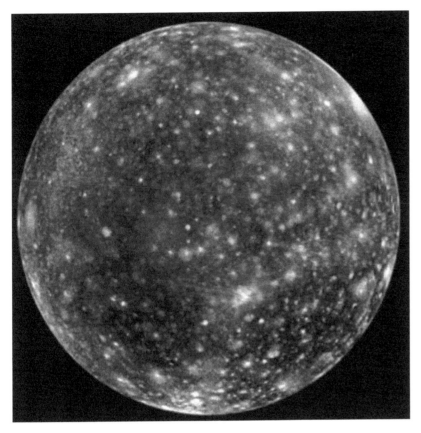

FIGURE 4.19 Callisto as imaged by the Galileo spacecraft showing the heavily cratered ancient surface of that moon. Bright spots are relatively recent impacts that have excavated Callisto's icy crust.

NASA/NSSDC

escaped to space by now, leaving Ganymede and Callisto cold and geologically dead.

This expectation is confirmed in the case of Callisto, whose surface is heavily scarred with meteorite craters, indicating, as in the case of our Moon, an ancient surface (Figure 4.19). Planetary surfaces accumulate impact craters over time, so that the older the surface, the more craters it shows.

Ganymede, however, seems to be geologically active, with a surface marked by many strange grooves and light-colored streaks (Figure 4.20). The cause of the geological activity and strange surface markings on Ganymede is uncertain. Some scientists have recently suggested that a thin ocean of liquid water might exist under a thick layer of surface ice on Ganymede.

4.4.3 Saturn

The second giant planet is Saturn (Figure 4.21). It too is composed mainly of hydrogen and helium. Saturn's average

FIGURE 4.20 Ganymede as imaged by the Galileo spacecraft shows heavily cratered dark areas and lighter bands of grooved and less cratered topography.
  NASA/JPL-Caltech.

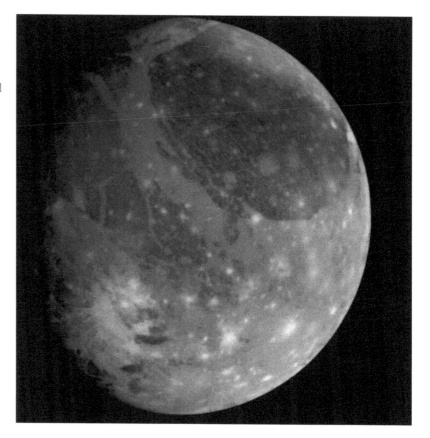

FIGURE 4.21 Saturn. A Hubble Space Telescope false color image of the planet. Saturn is about 75 000 miles (120 000 kilometers) in diameter at the equator, and the rings are about 100 000 miles (160 000 kilometers) in diameter.
  Erich Karkoschka, University of Arizona, and NASA

distance from the Sun is 886 million miles (1.42 billion kilo-meters) – nearly twice that of Jupiter. It completes a revolution around the Sun in approximately 30 years. The length of a day on Saturn is approximately 10 hours. Saturn is the second largest planet in the Solar System, being five-sixths the diameter of Jupiter and about a third as massive.

4.4.4 The moons of Saturn

Saturn has dozens of small moons and one large one. The small moons have densities only moderately greater than water, indicating that they are composed largely of ice plus smaller amounts of rock. The largest moon of Saturn, named Titan, has a diameter of 3180 miles (5088 kilometers) – larger than the planet Mercury (Figure 4.22). Its density – 1.9 grams per cubic centimeter – suggests a 50–50 mixture of rock and ice or water. Titan is unique among the moons of the Solar System because it possesses its own atmosphere.

Titan's thick atmosphere and heavy cloud cover prevent us from observing conditions on its surface, but observations from the Cassini–Huygens mission in 2004–5 show a surface temperature of −180°C. The atmosphere consists mainly of molecular nitrogen (N_2), but there is also about 5% methane (CH_4) and other hydrocarbons. Photochemical reactions in the atmosphere produce a rich variety of hydrocarbons, which form the haze that obscures the surface. Substantial amounts of hydrocarbons may have precipitated from the atmosphere onto the surface of Titan.

The Cassini–Huygens mission provided information on conditions on Titan's surface. In places there are large volcanoes made of frozen methane and water ice, suggesting that methane is still being outgassed from the interior of Titan (Figure 4.23). Features produced by tectonic activity, wind, and erosion and deposition by a fluid substance were also observed. The paucity of visible impact craters suggests that most craters were destroyed by geological activity, eroded away, or buried.

Some calculations and recent radar observations from Earth indicate that the gases condense into liquids, with pools of liquid methane and ethane on the surface, and occasional showers of ethane rain (Figure 4.24).

During the last part of the Cassini–Huygens mission, the Huygens probe landed on Titan and encountered a surface with a thin solid crust overlying soft material, interpreted as organic compounds precipitated from the atmosphere. It is interesting to note that this planet-sized moon has an abundance of the compounds – methane, hydrocarbons, and water – that can yield molecules important in the chemistry of life.

FIGURE 4.22 Titan. This view of Titan was taken by the Voyager 2 spacecraft from a distance of 2.8 million miles (4.5 million kilometers). Hardly any features are visible in the thick, unbroken haze that surrounds this large satellite. The brown color suggests the presence of compounds of carbon, nitrogen, and hydrogen.

NASA/NSSDC

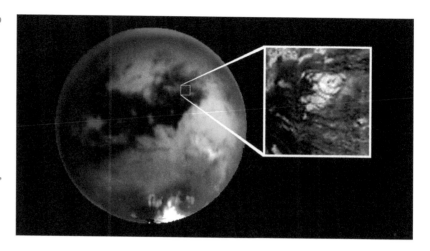

FIGURE 4.23 An ice volcano as seen in infrared light through the hazy atmosphere on Saturn's moon Titan. Since Titan's surface temperature is around −180 °C, lava welling up to form the volcanic mound could be a slurry of methane, ammonia, and water ice combined with other ices and hydrocarbons. The circular feature is roughly 19 miles (30 kilometers) in diameter.
NASA/JPL/University of Arizona

4.4.5 The rings of Saturn

The rings of Saturn are an extraordinary phenomenon familiar to every amateur astronomer. The four distinct rings visible from Earth occupy the region between 46 000 miles and 85 000 miles (74 000 and 136 000 kilometers) from the center of Saturn. The rings are very thin relative to their width. Consequently, when the rings are viewed edge-on to Earth, they disappear, although when viewed face-on they are dazzling bright.

Spacecraft photographs have revealed that the ring system of Saturn consists of many closely spaced ringlets (Figure 4.25). Stars can be seen through the rings, indicating that they are not solid sheets. Spacecraft observations have confirmed that the rings are made of particles ranging up to about 6 miles (10 kilometers) in diameter, and are probably composed of water and other ices.

The origin of the rings lies in the gravitational force exerted by Saturn on its satellites. Since the force of gravity increases in strength with decreasing distance, the near side of a moon feels a somewhat stronger gravitational pull toward its parent planet than the far side. The stronger attraction on the near side tends to wrench the material on this side out of the body of the moon. The force of the moon's internal gravity, holding it together, resists this effect. If the moon is too close to its planet, the excess pull on the near side becomes too great to be counteracted and the moon is torn apart. Each of the resulting fragments then circles in orbit around the planet as a miniature moon.

A moon must keep a minimum distance from its parent planet in order to stay intact. The minimum distance is called the *Roche limit*. A calculation shows that the rings of Saturn are inside the Roche limit for that planet. Either the rings are fragments of a moon that spiraled in too close and was torn apart; or, more likely, they are grains of material that were within the Roche limit around Saturn originally and, therefore, were prevented from collecting into a single object when the moons of Saturn were first forming. Recently, it has been proposed that icy comets approaching Saturn too closely were broken apart to form the ring material.

All planets and stars possess Roche limits. The Roche limit for the Sun is 1 million miles (1.6 million kilometers) from its center, or approximately 500 000 miles (800 000 kilometers) above its surface. No planet could form or remain within this distance. The Roche limit for Earth is 10 000 miles (16 000 kilometers) from its surface, well inside the orbit of the Moon.

4.4.6 Uranus and Neptune

Uranus and Neptune, the seventh and eighth planets from the Sun, were probably formed by a combination of accretion and gravitational condensation, in the same manner as Jupiter and Saturn. In spite of their smaller size and lesser degree of gravitational compaction, they have higher densities than Jupiter and Saturn, indicating that they also accreted a considerable amount of water and the heavier elements. Uranus has the unusual property that its axis of rotation is tilted approximately 90° to the plane of its orbit, so that the planet rotates, so to speak, on its side (Figure 4.26).

In 1986, the Voyager 2 spacecraft observed Uranus, its rings, and its five largest satellites at close range (ten small moons were also discovered). The innermost layers of the planet itself remain shrouded by its thick atmosphere of hydrogen and helium, with small amounts of methane and ammonia. Winds of more than 150 miles (240 kilometers) an hour have been measured, and by tracking cloud patterns in the upper atmosphere, scientists have calculated a period of

FIGURE 4.24 Radar image from the Cassini spacecraft showing a region that may be a shoreline. The light-colored region at the bottom appears to have channels cut by a moving fluid, while the smoother dark areas at the top may outline bays. Results from the Huygens probe that landed on Titan suggest that liquid methane might occupy some of these channels and bays intermittently. This radar image spans about 125 miles (200 kilometers).
NASA/JPL

FIGURE 4.25 A computer-enhanced photograph of the rings of Saturn, taken by the Voyager 2 spacecraft from a distance of 5 million miles (8 million kilometers). Computer processing has exaggerated the subtle color variations in this Voyager 2 view of the sunlit side of Saturn's rings.

NASA/NSSDC

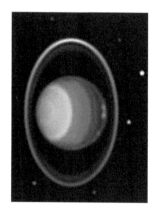

FIGURE 4.26 Uranus as seen by the Hubble Space Telescope.

Erich Karkoschka, University of Arizona, and NASA

rotation for the planet of about 16.8 hours. Average atmospheric temperatures may be as low as −250 °C (−420 °F). The thin rings of Uranus are complex and are believed to be composed mainly of boulder-sized objects, with only relatively small amounts of fine dust.

Uranus and Neptune have a bluish-green color that indicates the presence of methane gas, and clearly visible cloud belts and giant storms (Figures 4.26 and 4.27). The wind speeds on Neptune are higher than those on Uranus. The atmosphere of Neptune, like that of Uranus, is composed mainly of hydrogen and helium, with lesser amounts of methane. Neptune has a number of icy and rocky moons, and a system of thin dark rings.

4.4.7 Pluto

Pluto was the ninth and last planet to be found in the Solar System. It was discovered in 1930. The elliptical orbit of Pluto carries it farther from the Sun than that of any other planet – about 40 times the Earth–Sun distance – orbiting with a 248-year period. The planet is approximately 1440 miles (2300 kilometers) in diameter with a density of about 2.0 grams per cubic centimeter. Observations in 1978 revealed that Pluto itself has a moon, which has been named Charon,

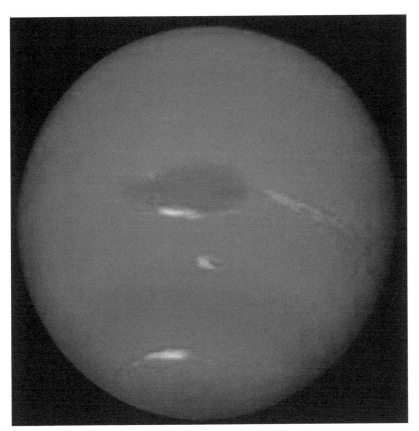

FIGURE 4.27 Neptune. The dark-blue and light-blue areas are the belts of clouds at various altitudes in the atmosphere of Neptune. The Great Dark Spot near the equator is a storm center more than 6000 miles (10 000 kilometers) in diameter.

NASA (NASA-HQ-GRIN)

about 740 miles (1186 kilometers) in diameter (Figure 4.28). The large size of Charon relative to Pluto makes the pair a "double planet." Estimates of the density of the two bodies suggest that Pluto and Charon are composed mainly of a mixture of ice and rock. There is direct evidence that at least part of Pluto's surface consists of frozen methane, and the planet may also have a thin methane atmosphere.

Recently, there has been a move to demote Pluto from its position as a planet. Since the 1980s, a number of icy bodies up to hundreds of miles in diameter have been found in orbits similar to Pluto's. These objects define what is known as the *Kuiper Belt* (or Edgeworth–Kuiper Belt). Recent work suggests that some 70 000 objects with diameters greater than 50 miles (80 kilometers) orbit the Sun at distances of 30 to 50 times the Earth–Sun distance. The total number of Kuiper Belt objects of all sizes probably numbers in the billions. In this view, Pluto would be the largest of the Kuiper Belt objects.

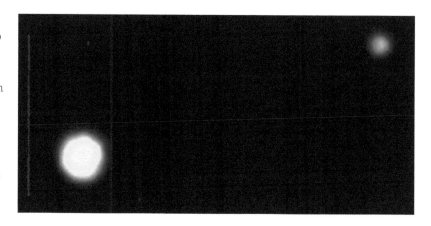

FIGURE 4.28 Pluto and Charon. This image of Pluto and its moon Charon was taken by the Hubble Space Telescope. Pluto and Charon are separated by only about 12 000 miles (19 000 kilometers).

Rudi Albrecht, ESA FOC IDT, and NASA-ESA Hubble Space Telescope

4.5 Minor bodies of the Solar System

The Solar System contains a large number of smaller objects that are held in orbit by the Sun's gravity. These minor bodies of the Solar System – asteroids, meteoroids, and comets – are negligible in mass compared to the planets. However, they can create striking visual effects if they come close to Earth, and they can collide with the planets. They also have a special scientific interest because they provide information on the primordial state of condensed matter in the solar nebula before it aggregated into the full-sized planets.

4.5.1 Asteroids

Between the orbits of Mars and Jupiter there is a gap in the distribution of the planets. We might expect to find a planetary body located outside the orbit of Mars, about three times Earth's distance from the Sun; but instead we find only a large number of small, mostly rocky and metallic bodies – planetesimals – circling in a ring from about 1.8 to 4 times the Earth–Sun distance. These are called *asteroids* (Figure 4.29). The largest of the known asteroids is Ceres with a diameter of 480 miles (770 kilometers), discovered in 1801. Three other asteroids – Pallas, Vesta, and Hygiea – have diameters greater than 200 miles (320 kilometers). Most of the remaining asteroids – estimated to number tens of thousands – are much smaller. At present, there are about 150 000 known asteroids with thousands of new ones discovered each month. All but the biggest asteroids were originally larger bodies that have been ground down to smaller size by repeated collisions with each other. That is why the present asteroids have very irregular shapes.

Asteroids whose orbits take them inside the orbit of Earth are called Earth-crossing, or Apollo asteroids. These asteroids were perturbed from their normal orbits within the asteroid belt by Jupiter's gravity. Currently, more than 2500 Earth-crossing asteroids are known, with about 700 of those larger than 1 kilometer in diameter. Some Apollo asteroids come dangerously close to our planet (Figure 4.30). The impact of a large asteroid would liberate the energy of millions of hydrogen bombs, and might destroy a substantial fraction of life on Earth. We know from the geological record that Earth has been struck by a number of large Apollo objects during the last few hundred million years. Sixty-five million years ago, the impact of a 6-mile (10-kilometer) diameter asteroid led to the extinction of the dinosaurs, and 75% of all species living on Earth at the time.

FIGURE 4.29 Galileo spacecraft image of the asteroid 243 Ida (longest dimension equals 35 miles or 56 kilometers) and its moon, Dactyl (diameter about 1 mile or 1.6 kilometers). Like Ida, Dactyl has a heavily cratered surface. It is probably a piece of Ida that was ejected from the asteroid when it was struck by another asteroid fragment.

NASA Jet Propulsion Laboratory (NASA-JPL)

4.5.2 Meteorites

Collisions between asteroids produce fragments that can be ejected from the asteroid belt. Some fall by chance into orbits crossing the orbit of Earth. It is believed that most of the meteorites that hit Earth have this origin. (The term *meteroid* is used when those objects are in space, and the term *meteorite* is used after they land on Earth.) The examination of meteorites that survive the passage through Earth's atmosphere reveals that most are pieces of rock and iron with a rather complex physical and chemical history, suggesting that they were broken from larger bodies during repeated collisions.

Most meteorites consist primarily of rocky materials (silicate minerals), but fragments of metallic iron and nickel are not uncommon (Figure 4.31). The meteorite that blasted out Meteor Crater in Arizona was a large chunk of iron/nickel weighing approximately 1 million tons (1 billion kilograms).

Meteorites range in size from blocks of material weighing many tons to nearly invisible grains of rock and iron called micrometeorites. If the meteorite is the size of a grain of sand or larger, it leaves a fiery trail of incandescent matter behind as it passes through Earth's atmosphere. When we see this

FIGURE 4.30 The orbits of various asteroids. Most asteroids orbit the Sun in a belt between the orbits of Mars and Jupiter. The large asteroids 1 Ceres, 2 Pallas, and 3 Juno all have roughly circular orbits that lie within this belt. By contrast, some asteroids, such as 1862 Apollo and 1566 Icarus, have eccentric orbits that carry them inside the orbit of Earth.

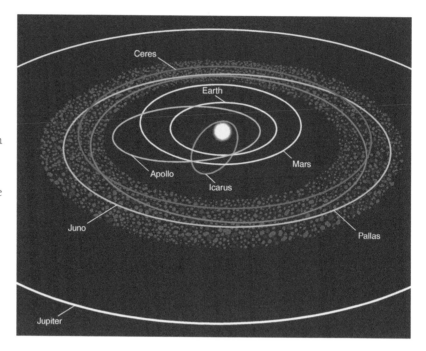

trail in the night sky, we call it a *shooting star* or *meteor*. The great majority of such objects are so small that the heat of their passage through the atmosphere vaporizes them before they reach the ground. If the meteorite is the size of a basketball or larger, its entry into the atmosphere creates a spectacular sight called a fireball.

Prior to astronauts landing on the Moon, meteorites were the only samples of extraterrestrial matter available for study. Their examination in the laboratory has provided two results of great importance to students of the origin of the Solar System. One relates to the age of the Solar System. Measurements of the ages of meteorites by the technique of radioactive dating yield results ranging up to 4.6 billion years. These 4.6-billion-year-old meteorites are the oldest objects in the Solar System and can be used to estimate the age of Earth. This conclusion is strengthened by the dating of some lunar rocks, which also have ages up to about 4.6 billion years.

A second point of interest is the composition of meteorites. Most meteorites show little sign of having been worked over by air and water erosion or subjected to a long history of chemical separation, as is the case of rocks on the surface of

Earth. For this reason, they are believed to give a better indication of the original state of the materials in the solar nebula than can be obtained on Earth. Two relatively rare classes of meteorites have proven very valuable for information on the planets. These are meteorites derived from the Moon and Mars, blasted off those bodies by large asteroid impacts and eventually landing on Earth.

4.5.3 Comets

Comets are snowball-like mixtures of ice and rock that originated during the formation of the Solar System 4.6 billion years ago. Frozen gases in comets include water (H_2O), carbon dioxide (CO_2), carbon monoxide (CO), and possibly methane (CH_4). The comet appears first in the telescope as a small, fuzzy, faintly luminous object. At this point it is far from Earth, but is approaching our planet on the inward leg of a long journey from the edge of the Solar System. As it comes closer, solar energy warms the head of the comet and vaporizes gases that were frozen during the many years in which the comet was far from the Sun. These gases stream out behind the comet's head. Excited to luminescence by the absorption of solar radiation, the stream of gases forms a spectacular, glowing tail, which becomes clearly visible to the naked eye as the comet nears the Sun (Figure 4.32). As the comet recedes from the Sun, the tail disappears again. Eventually, after many orbits, a comet may lose most or all of its volatile constituents.

Comets derive their name from the appearance of the comet tail; since the word comes from the Latin *cometes*, which means "long-haired." All comet orbits are elliptical. Most, however, are highly elongated, approaching closely to the Sun, crossing the orbits of the planets on their approaches to the Sun, and then retreating far beyond the orbit of Pluto on the outward leg of their journeys. Many comet orbits cross the orbit of Earth. The most elongated orbits are estimated to reach one-fifth of the way to the next nearest star. A comet in one of these orbits requires several million years to complete one circuit of the Sun, and most comets are estimated to require 10 000 years or more for the round trip. Thus, the vast

FIGURE 4.31 Two common types of meteorites. (Top) An iron meteorite composed almost entirely of the elements iron and nickel. (Bottom) A stony meteorite. The surface of a typical iron meteorite is covered with thumbprint-like depressions caused by ablation (removal by rapid melting) during the meteorite's rapid descent through the atmosphere.

FIGURE 4.32 This image shows Comet Hale–Bopp on March 8, 1997. The heat of the Sun causes the comet's nucleus to evaporate, forming a white tail of dust particles and a bluish tail of ionized atoms and molecules. Alan Hale and Thomas Bopp discovered this comet independently on the same night in July 1995.
NASA Jet Propulsion Laboratory (NASA-JPL)

majority of the comets we observe will not be seen again from Earth for thousands of years. Trillions of these bodies reside in the outer fringes of the Solar System, in a vast spherical cluster called the Oort Cloud, located at a distance of about a light-year from the Sun. Recent findings suggest that they were ejected from the zone of the outer planets.

Some comets have shorter periods, ranging down to 3.3 years for Comet Enke. Most of these short-period comets come close to the orbit of Jupiter, suggesting that Jupiter's gravity has deflected them into new orbits that remain within the Solar System. Apparently, many comets collide with Jupiter and the other giant planets. In 1994, Comet Shoemaker–Levy 9 passed so close to Jupiter that it was broken by gravitational forces into a number of fragments that crashed into Jupiter on the comet's subsequent orbit around the giant planet (Figure 4.33).

The most famous of the short-period comets is Comet Halley, named after Edmond Halley, a contemporary of Newton, who studied the records of comets dating back to 1531 and concluded that several of these comets were a single body making a repeated appearance in the sky. Halley predicted that this comet would return in 1758, and it reappeared on Christmas night of that year. The period of Halley's Comet varies from 74 to 79 years from one orbit

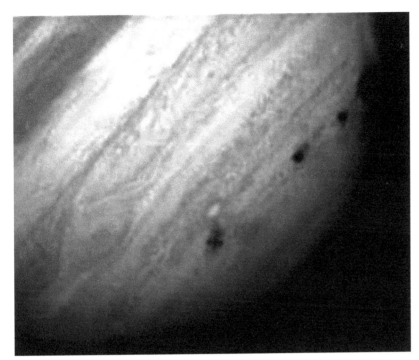

FIGURE 4.33 Hubble Space Telescope image of Jupiter in July 1994 showing the sites of impact of several fragments of Comet Shoemaker–Levy 9. The dark spots are enormous plumes of hot gases in Jupiter's thick atmosphere rising above the impact sites.

Hubble Space Telescope Comet Team and NASA

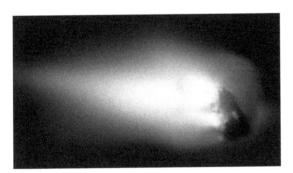

FIGURE 4.34 The nucleus of Comet Halley. This close-up picture, taken at a distance of about 373 miles (597 kilometers) by a camera onboard the European Space Agency spacecraft Giotto in March 1986, shows the nucleus of the comet. The nucleus is darker than coal and measures 9 miles (14 kilometers) in the longest dimension. Two bright jets of dust extend from the nucleus in the direction of the Sun (to the left in the figure).

ESA. Courtesy MPAe, Lindau

to the next as the result of changes in the comet's orbit produced by Jupiter's gravity. The nucleus of Comet Halley is about 9 miles (14 kilometers) in diameter, with a very loose structure and a low average density. In 2001, the Deep Space 1 spacecraft passed within 1400 miles (2240 kilometers) of Comet Borrelly, which is believed to have originated in the Kuiper Belt near Pluto's orbit. The comet's nucleus is about 5 miles (8 kilometers) long and shaped like a bowling pin. The surface of the comets is covered with a dark carbon-rich crust similar to that found on Comet Halley in a similar flyby in 1986 (Figure 4.34).

Two missions to comets are under way. The NASA Deep Impact mission and the Japanese Hayabusa sampler involve encounters with comets in order to measure their physical and chemical properties. The idea is to crash large probes (in the case of Deep Impact) or fire bullets (in the case of Hayabusa) into the surface of two comets, and capture some of the material that flies up from the blast. The effects of the blasts should tell us the physical makeup of the comets, and the samples when returned to Earth will tell us the chemical composition of these icy objects.

4.6 Summary

By the end of this chapter you should be familiar with the following concepts and topics:

Formation of the planets in our own Solar System
The rotating cloud of gas and dust
Small terrestrial planets accreted from rocky material in
   the warm inner regions
   Mercury
   Venus
   The Earth–Moon system
   Mars
Giant gaseous planets and their satellites formed in the
   cold outer regions
   Jupiter
   The Galilean satellites
   Saturn
   The moons of Saturn
   The rings of Saturn
   Uranus and Neptune
Pluto
Minor bodies of the Solar System
   Asteroids
   Meteoroids
   Comets

Questions

4.1. List the distances of the planets from the Sun in AU.

4.2. What general properties of the Solar System are consistent with origin from a rotating accretion disk?

4.3. How did the terrestrial and giant planets form?

4.4. What are Saturn's rings made of?

4.5. What are the moons of the giant planets like?

4.6. Where and what is the asteroid belt? What hypotheses are offered for its origin?

4.7. What are comets made of, and where do they reside in the Solar System?

4.8. What are meteorites? How old are they?

4.9. What does the current theory of the origin of planets suggest regarding the likelihood of other solar systems in the Universe? Describe the recent evidence for other planetary systems.

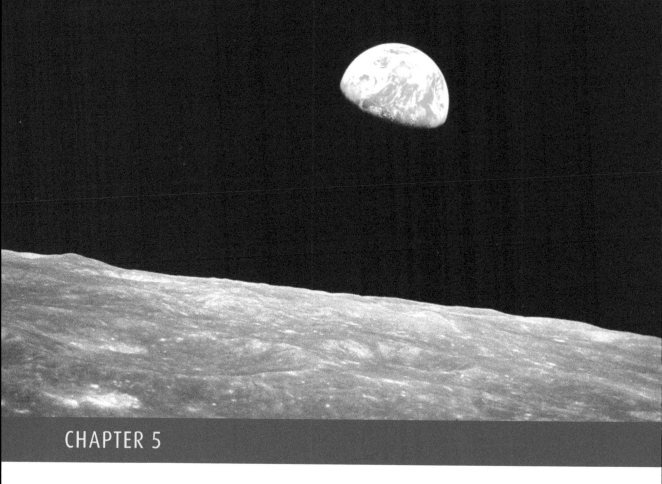

CHAPTER 5

Origins of a habitable planet

The Moon as a record of early Earth history

Earth has not always been the oasis of life it is today. Yet the history of the first 700 million years on Earth has been entirely wiped away by natural forces such as the air and water that now make our planet habitable. Before we embark on a detailed study of our planet, we must first look at the properties of planetary bodies similar to Earth – the Moon, Mars, and Venus. Their properties tell us some important things about Earth and its origin that we could not learn from the study of our planet alone. The Moon has turned out to be a Rosetta Stone for deciphering the early history of the terrestrial planets. The bleak and inhospitable conditions that prevail on the Moon have preserved the evidence of this history that has been destroyed by geological activity on Earth.

On the Moon there are no oceans and atmosphere to rework the surface, and little of the geological activity that rapidly changes the face of the Earth. This chapter takes a closer look at the Moon, and shows how astronomers have been able to reconstruct its full life story using results from the Apollo missions and subsequent lunar observations. We will learn how rock samples brought back to Earth indicate that the surface of the Moon was once entirely molten, and how this could have happened. We will also look at the origin of the Moon, and why it evolved into a world so different from Earth.

5.1 The Moon as a Rosetta Stone

Earth and the Moon, along with the rest of the Solar System, formed about 4.6 billion years ago. Earthbound creatures can never discover the conditions that prevailed on our newborn planet, because the record of Earth's early years has been largely wiped out. The air and water that make our planet livable have worn down the oldest rocks and washed away their remains into the oceans, while mountain-building activity and volcanic eruptions have churned the surface and flooded it repeatedly with fresh lava, removing the remaining evidence. These natural forces have entirely removed the primitive materials that lay on Earth's surface when it was first formed. No rocks have ever been found on Earth that are older than 3.9 billion years.[1] We know little of what happened on our planet from the time of its formation, 4.6 billion years ago, to the time when these oldest rocks were laid down. The first 700 million years are mostly blank pages in Earth's history.

But on the Moon there are no oceans and atmosphere to destroy and rework the surface and there is little of the geological activity that rapidly changes the face of Earth. Over large areas, the materials of the Moon's surface are as well preserved as if they had been in cold storage. The Moon offers

---

[1] Grains of the mineral zircon extracted from some of the oldest rocks on Earth have radiometric ages of up to 4.4 billion years. These grains were derived from older rocks that otherwise have been completely destroyed.

Previous page: Earthrise on the Moon. Taken July 20, 1969 as the crew of Apollo 11 returned from the far side of the Moon. NASA

the best chance of recapturing the lost record of Earth's early history. It is Earth's Rosetta Stone.

5.2 The surface of the Moon

When Galileo, the first person to look at the Moon through a telescope, turned his primitive instrument on that body in 1609, he saw large, dark areas resembling Earth's oceans, and mountainous light-colored areas that seemed to resemble the continents. This pattern of light and dark regions, visible to the naked eye, makes up the face of the man-in-the-Moon (Figure 5.1). Galileo thought the dark areas were actually oceans, and called them *maria*, or seas (Figure 5.2). The light-colored, mountainous regions came to be known as the lunar *highlands* (Figure 5.3).

FIGURE 5.1 The ancient surface of the Moon, showing the dark seas or maria and the bright heavily cratered highlands.
   USGS

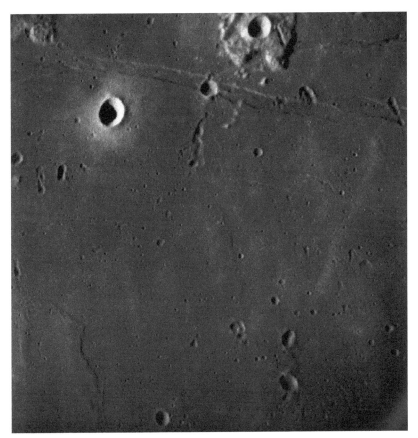

FIGURE 5.2 A mare region on the Moon. This photograph shows the Sea of Tranquility near the Apollo 11 landing site. The prominent linear feature at the top of the photograph is Hypatia Rille, a fault valley or collapsed lava tube. The prominent crater just below Hypatia Rille is Moltke.

© Lunar and Planetary Institute, 2007

FIGURE 5.3 The heavily cratered lunar highlands of the Moon. This Earth-based photograph shows a portion of the Moon's southern hemisphere about 370 miles (600 kilometers) across.

From the Consolidated Lunar Alas © Lunar and Planetary Institute, Universities Space Research Association

Today we know that these similarities to the surface of Earth are illusory. The lunar seas contain no water; no storms rage across the dark plains; no streams flow down from the highlands. And the lunar highlands are not similar in any way to Earth's continents; they do not resemble continental rocks chemically, and the forces that created the lunar highlands were different from the forces that thrust up the great mountain ranges of Earth.

A casual inspection of the Moon through a telescope confirms that the Moon's surface is completely different from that of Earth. Photographs of the Moon taken through a large telescope show that the lunar highlands are pitted by innumerable craters of all sizes. Most of these craters were produced by the impact of meteorites that have been raining down on the Moon's surface for billions of years. Many large craters are circled by ramparts ranging up to 10 000 feet (3000 meters) in height. Many of these ramparts are more than 1 billion years old, yet photographs taken with a telescope clearly indicate that they have been preserved almost unchanged, with little of the original material worn away (Figure 5.3).

Meteorites have collided with Earth throughout its history, just as they have collided with the Moon, and they have produced similar craters; but most traces of the older craters are highly eroded. Only the scars of relatively recent collisions, such as Meteor Crater in Arizona, formed about 50 000 years ago, are still well preserved on Earth (Figure 5.4).

Pictures of the Moon taken by NASA spacecraft provided evidence that the extent of erosion on the Moon has indeed been very small. The photographs showed lunar features as small as a meter in diameter. The clarity of the spacecraft photographs produced jubilation among astronomers, who had been straining to see through Earth's atmosphere like drivers peering at the road through a rain-spattered windshield. Many craters that had never been seen before in photographs taken with telescopes on Earth were visible, ranging from about a meter (3 feet) in diameter up to hundreds of meters. These small craters must have existed on Earth as well, but were wiped out almost immediately by the weathering effect of winds and running water. On the Moon, such craters can last for a billion years, and even the shallow

FIGURE 5.4 Meteor Crater in Arizona. An iron meteoroid measuring about 150 feet (50 meters) across struck the ground in Arizona about 50 000 years ago. The result was this impact crater, about 0.8 miles (1.2 kilometers) in diameter and 600 feet (200 meters) deep at its center.

D. Roddy, Meteor Crater Enterprises

footprints of the Apollo astronauts, no more than a few centimeters deep, will still be visible millions of years from now (Figure 5.5).

The Moon's resistance to the ravages of time endows its barren landscape with a unique value. The rocks that litter the Moon's surface contain no life; we know they contain very little gold or silver; nonetheless, they are scientifically price-less because of the revelations they can offer regarding the early years of the Solar System.

Sometimes lunar geologists refer to events occurring on the Moon several million years ago as "recent" events. They are recent in the sense that they occurred only a short time ago in comparison to the age of the Moon and the age of the Solar System. But consider how the surface of Earth has been trans-formed in, say, the last 200 million years. Two hundred mil-lion years ago the continents of Earth were located in entirely different places than they are today: Africa and South America were joined in a super-continent; the Rocky Mountains, the Alps, and the Himalayas did not exist; and the eastern sea-board and most of the southwest region of the United States were at the bottom of shallow seas. Yet the Moon has scarcely changed at all in this period.

Because of these circumstances, scientists looked forward with considerable interest to the exploration of the Moon

FIGURE 5.5 Buzz Aldrin's footprint on the surface of the Moon (from the Apollo 11 mission). The fine dust that covers the Moon is produced by the countless impacts that strike its surface.

NASA.

FIGURE 5.6 This view of the Moon's surface shows the lunar roving vehicle near the landing site for Apollo 17. Astronauts Harrison Schmidt and Eugene Cernan are sampling unusual "orange soil" at this locality at the eastern edge of the Sea of Serenity.

NASA

during the Apollo lunar landings. The Apollo landings yielded a treasure trove of 836 pounds (about 400 kilograms) of rocks, a huge amount of scientific data, and many thousands of photographs (Figure 5.6).

5.3 The Apollo findings

The chemical ingredients of the Moon's surface have been analyzed with the aid of instruments carried around the Moon in orbit during the Apollo flights and on later space missions, and by studying rock samples in the laboratory. The results show that the entire surface of the Moon – including the maria and the highlands – appears to be covered with igneous rocks – a type of rock formed by the cooling of molten material. Thus, this single finding from Apollo indicates that the surface of the Moon was entirely molten at some point in its past.

What melted the Moon's surface? Only two possible answers are known. One is that the Moon was melted by an intense meteorite bombardment during a short period at the beginning of its life. This explanation fits in with the currently favored theory on the origin of the Moon, which proposes that the Moon accreted out of particles of dust and fragments of rock of various sizes. As our Moon grew to final size in the last stages of this birth process, gravity pulled the material around it down onto the surface with great force. Each rock generated some heat as it crashed into the surface. Normally, this heat would be radiated away slowly to space, but if enough impacts occurred in a short period of time, the total accumulation of heat could be sufficient to melt the Moon to a considerable depth.

The alternative explanation is that the Moon was melted by radioactive energy released in the decay of uranium and other radioactive elements scattered throughout the Moon's interior. However, radioactive elements release their heat very slowly. In fact, calculations show that it takes about 1 billion years of steady radioactive heating to bring the interior of a body like the Moon or Earth to the melting point. If the rocks on the surface of the Moon were melted when the Moon was roughly 1 billion years old, that fact would suggest that radioactive heat was the likely cause of the melting; but if the rocks were melted when the Moon was considerably younger than 1 billion years, it would be necessary to look to another factor – presumably meteorite bombardment – for the explanation.

5.3.1 Ages of the lunar rocks

These ideas indicate that the times at which the Moon rocks were melted could provide the clue to the cause of their melting. With this remark we come to the second critical Apollo result, which is the measurement of the ages of the Moon rocks. The ages of the rocks are measured by the technique of radioactive decay, which tells how long they have been in their present crystalline form. In other words, it tells how much time has elapsed since these rocks were last melted. Thus, the age measurements provide precisely the information needed to distinguish between the two causes of the Moon's melting.

But nature rarely gives up her secrets without a struggle. When the results of the age measurements became available, they yielded the ambiguous answer that *both* causes of melting probably had played roles in the Moon's history. The lunar

FIGURE 5.7 Apollo 17 astronaut Harrison Schmidt sampling a huge split boulder ejected from a large impact crater on the Moon.
NASA

highland rocks were melted by meteorite bombardment early in the Moon's life, while the lunar seas, covered by the dark igneous rock basalt, were melted later by internal heat.

This fact did not become clear until the astronauts were in true highlands terrain. All the highland rocks collected during the Apollo missions turned out to be a fragmental rock called *breccia*. The fragments in the highland rocks were produced by massive bombardment by meteorites (Figure 5.7). Several highland rocks – the oldest rocks found on the Moon – gave ages of 4.44 billion years. From the age of these rock fragments it follows that the surface of the Moon must have been melted and re-solidified when the Moon was only about 100 million years old. One hundred million years is probably too short a time for the Moon to have been melted by the slow process of radioactive heating. Therefore, something else, such as meteorite bombardment, must have melted the highland rocks.

The highlands are known to be older than the lunar seas, and are thought to be derived from the Moon's original crust, parts of which were later covered over by the darker materials of the lunar maria. Thus, it appears that the original surface of the Moon was largely melted during the Moon's birth, or shortly after.

5.3.2 Evidence for internal melting

But the story of the Moon's melting does not end there, because other Moon rocks – those collected from the maria – have younger ages, ranging from 3.1 to 3.8 billion years. Since the Moon was formed 4.6 billion years ago, these ages imply that the materials of the maria began to melt when the Moon was about 800 million years old. This time period would correspond to the time required for radioactive heating of the Moon to melt the rocks in its interior, leading to floods of molten lava on the surface. The lunar seas are pools of solidified lava that accumulated in large basins in the original highland crust as a result of repeated volcanic eruptions. The huge basins themselves were apparently created by giant impacts during a particularly intense period of bombardment about 3.9 to 3.8 billion years ago.

Another important conclusion follows from the fact that the oldest mare rocks are 3.8 billion years old and the youngest go back 3.1 billion years. This spread of ages suggests that the Moon experienced an episode of sustained volcanism lasting about 700 million years. The 700-million-year interval of melting and lava flooding on the Moon was undoubtedly accompanied by emissions of gases such as carbon dioxide and water vapor and all the other manifestations of internal heat within a planet that are so familiar to us through our experience with volcanic activity on Earth. When the volcanism ended, atmospheric gases were no longer released from its interior and the Moon subsided into geological lifelessness. For the last 3 billion years, the Moon has been geologically dead.

5.4 The origin of the Moon

Armed with the facts from Apollo and subsequent lunar observations, we can reconstruct the full life story of the Moon. Until recently, however, the question of the Moon's

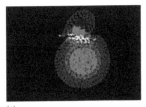
(a)

(b)

(c)

(d)

(e)

(f)

FIGURE 5.8 The giant impact theory of the Moon's origin. According to this theory, a body the size of Mars collided with Earth about 4.5 billion years ago. This computer simulation shows the creation of the Moon from material ejected in this collision. Blue and green indicate iron from the cores of Earth and the asteroid; red, orange, and yellow indicate rocky mantle material. The impact ejects both mantle and core material, but most of the iron falls back onto Earth. The surviving ejected rocky matter coalesces to form the Moon (indicated by arrow) during its first orbit of Earth.

Willy Benz, Physikalisches Institut, Universität Bern, Switzerland

origin had been a longstanding problem for lunar scientists. Several competing theories argued that Earth and the Moon were formed separately, that the Moon separated from Earth early in its history, or that the Moon was somehow captured by Earth.

A clue to the Moon's origin is the fact that it is generally depleted in iron and similar elements relative to Earth. The mean density of the Moon is only about 3.3 grams per cubic centimeter as compared with 5.5 grams per cubic centimeter for Earth. This finding, along with measurements of the moment of inertia of the Moon, tells us that the Moon cannot have very much high-density material at its center, and hence lacks a significant iron–nickel core. The Moon is also depleted in relatively volatile elements such as sodium, which could have been lost if the materials that built the Moon were violently heated.

The depletion of iron and similar elements in the Moon rocks, the apparent absence of a large iron–nickel core within the Moon, and the great depletion of volatile elements, are all in line with a catastrophic origin of the Moon from material ejected from the outer rocky parts of Earth after most of the iron and related elements were sequestered in Earth's core.

The evidence suggests that our satellite was born in the collision between Earth and a Mars-sized body. The Moon accreted out of the remains of the huge asteroid and material thrown out of Earth during that catastrophic collision (Figure 5.8). The ejected material initially formed a ring around Earth, and eventually coalesced to form the Moon. As the Moon grew in size, its surface was bombarded by a vast quantity of meteorites which rained down with ever increasing speed. The rocks in each collision area had no chance to cool off from the heat of one impact before they were warmed up by the next. The temperature rose as the intense bombardment continued. Eventually, the outer layer of the Moon melted, and remained molten until the intense bombardment subsided.

Why did volcanic activity die out on the Moon about 3 billion years ago? The answer probably is that the Moon, being a small planet, lost its internal heat to space rapidly. The Moon's diameter of 2100 miles (3400 kilometers) is only

FIGURE 5.9 Earth and its Moon to scale. While on its way to Jupiter, the Galileo spacecraft recorded this image of Earth (at lower right) and the Moon (at upper left) from a distance of about 3.9 million miles (6.2 million kilometers) on December 16, 1992.
NASA/NSSDC

about one-quarter of Earth's diameter, and its mass is only about 1/80 that of Earth (Figure 5.9). As the Moon's heat dissipated, and its internal temperatures dropped, the extent of the warm, molten region diminished, and the outer zone of cold, strong rock became thicker. According to data from seismic studies of the Moon, this cold outer zone today extends from the surface to a depth of 500 miles (800 kilometers). Beneath this thick layer, there exists a warmer zone of rock; below a depth of 600 miles (960 kilometers) the Moon may be partly molten.

It is extremely unlikely that molten rock lying at such great depths could force its way upward through the thick layer of solid rock that caps the Moon. This casing of cold rock probably explains the absence of volcanic eruptions on the surface of the Moon during the last 3 billion years. This rigid outer casing of rigid rock is also too thick to break up into great slabs that move against each other, as Earth's surface does, producing earthquakes, volcanoes, and mountain building.

These circumstances account for much of the difference between the geology of the Moon and the geology of Earth, and explain why the appearance of the Moon today is so different from that of our own planet. Had the Moon been larger and thus retained its internal heat and volcanic activity for a longer period and a significant amount of atmospheric gases could have accumulated, Earth's sister body might be a more habitable place today.

5.5 Summary

By the end of this chapter you should be familiar with the following concepts and topics:

The full life story of the Moon
Lunar observations
The Apollo missions
  Ages of the lunar rocks
  Evidence for internal melting
The origin of the Moon
Its evolution into a world so different from Earth

Questions

5.1. What are the sources of heat inside a planet?

5.2. What is the relationship between the Moon and Earth in size? What is the distance from Earth to the Moon?

5.3. What are the differences between the lunar highlands and the lunar maria?

5.4. What is the currently accepted view of the origin of the Moon?

5.5. How are impact craters produced? Why does the Moon have so many craters?

5.6. Why can it be said that the Moon is a Rosetta Stone in deciphering Earth's early history?

CHAPTER 6

Prospects for life

Is Mars a habitable planet?

Mars has been a major focus in the search for extraterrestrial life. After Earth, it is the planet with the most hospitable climate in our Solar System. Its remarkable geology is suggestive of a planet that once had flowing rivers, active volcanoes, and the potential for life to arise. Water is an essential ingredient for the development of life as we know it, and has become a key element in the discussion of the prospects for life on Mars, past and present. The planet has been intensively studied by spacecraft and from Earth, yet many questions remain. Why are the surface conditions on Mars so very different from those on Earth? Were the conditions necessary for life to originate ever present on Mars?

Many characteristics of Mars are similar to those of Earth, and suggest the possibility for life, either now or in the past. In this chapter we will see how spacecraft images have enabled us to build up a detailed understanding of the planet. We will learn why scientists think Mars once had an abundance of liquid water and a substantial atmosphere, and look at evidence of frozen water at the poles and beneath the surface of the planet today. In building up a picture of the history of water on Mars we will begin to understand why it has become such a focus for astrobiology and the search for life.

6.1 Mars

Mars has generated more speculation regarding extraterrestrial life than any other planet in the Solar System. Several characteristics of the planet have contributed to the growth of these speculations. Its surface reveals changes during the Martian year that resemble the march of the seasons on Earth. In each hemisphere, a polar cap grows larger in the fall and winter and diminishes in the spring and summer. Dark regions appear each spring that are suggestive of the seasonal growth of vegetation. Toward the end of the nineteenth century, some observers reported a planetary network of canals, presumably engineered by intelligent life.

Modern studies of Mars failed to confirm the existence of plant life or of the canals. However, Earth-based and spacecraft observations made in recent years have contributed a great deal of new information that bears directly on the prospects for Martian life.

Mars has a thin atmosphere, composed almost entirely of carbon dioxide, about a hundredth as dense as Earth's atmosphere. A trace of moisture exists in the atmosphere, equivalent to a film of water less than a thousandth of an inch thick. Although this small amount of water is not adequate to permit life to originate, it could support life that had evolved earlier, in a wetter and more favorable age, and had then adapted slowly to drier conditions. The suggestion that Mars once had an abundance of water has received support from spacecraft observations carried out in orbit above Mars and on

Previous page: The surface of Mars from the Viking 1 Lander.
NASA

its surface. All in all, as an abode of life, Mars is, next to Earth, the most promising planet in the Solar System.

6.1.1 Geology of Mars

Mars is approximately twice as large as the Moon and half the size of Earth by diameter. Its mass is about ten times the Moon's mass and one-tenth the mass of Earth. These facts lead to a prediction regarding volcanism and geological activity on Mars. Volcanism results from the release of heat within the body of an earthlike planet. The source of the heat is most likely the radioactive decay of the elements uranium, thorium, and potassium, which exist in the interior of an earthlike planet in small concentrations of a few parts per million. These elements exist in very roughly the same concentrations in the Moon and Earth. Presumably they also exist in the interior of Mars.

FIGURE 6.1 Global mosaic of Viking 1 Orbiter images of Mars. At center is Valles Marineris. At left are the Tharsis volcanoes and to the south is ancient, heavily cratered terrain.

NASA/NSSDC

The heat that radioactive elements release must have gradually increased the temperature in the interiors of the Moon, Mars, and Earth during their early years. Volcanism may have commenced on the Moon, Mars, and Earth at about the same time as the product of this chain of circumstances.

Although the volcanism probably began at about the same time on each planet, it did not last for the same length of time on each, because of the differences in their sizes. The reason is that the heat generated in the interior of a small planet has to travel only a short distance to reach the surface. Heat is lost from the surface of a small planet at a relatively rapid rate because of this fact. If the planet is very small, its temperature will never reach the melting point of rock, and volcanism will never commence.

In other words, small planets, like small animals, lose their internal heat at a relatively rapid rate. If a planet is somewhat larger, its internal temperature may reach the melting point of rock, but will not remain there for more than a short time. This appears to have been the case for the Moon, whose episode of volcanism lasted only about 700 million years.

If a planet is very large, its internal heat must travel through a relatively thick layer of rock to reach the surface. This thick layer acts as an insulating blanket, bottling up the radioactive heat, and causing the temperature in the interior to remain at a high level for a relatively long period of time. This is the case for Earth, which has remained volcanically active throughout most of its lifetime, and is still active.

This chain of reasoning leads to a prediction regarding the geology of Mars. This planet, intermediate in size between the Moon and Earth, must have retained its radioactive heat longer than the Moon, but not as long as Earth. Therefore, Mars should have been volcanically active for longer than 700 million years, and may have been active for as long as several billion years. This remark implies that Mars may have been the scene of active emission of volcanic gases until a relatively recent period in its geological history.

Water vapor and the other gases released by internal heating are the probable sources of water on Earth's surface and gaseous elements in its atmosphere. If volcanic activity persisted on Mars for two or three billion years, and the relative amount of water trapped in the interior of Mars was roughly the same as the amount of water trapped within Earth, the surface of Mars may once have been covered by water to a considerable depth. A substantial atmosphere also may have existed during this long period of volcanic activity.

6.1.2 Evidence for geological activity

The suggestion of volcanism on Mars was confirmed in 1971 when television cameras, mounted on a satellite placed in orbit around Mars, obtained excellent pictures of the entire planet. When the first pictures were obtained, the surface of the planet was obscured by a violent dust storm that persisted for nearly two months. Eventually the dust settled, and the television cameras revealed details of the surface of Mars that had never before been seen.

The most conspicuous feature was the huge volcanic mountain, Olympus Mons (Figure 6.2). The crater at the summit of this mountain is about 40 miles in diameter. The entire mountain is 300 miles (about 500 kilometers) across at its base, and rises at least 70 000 feet (21 000 meters) above the surrounding floor.

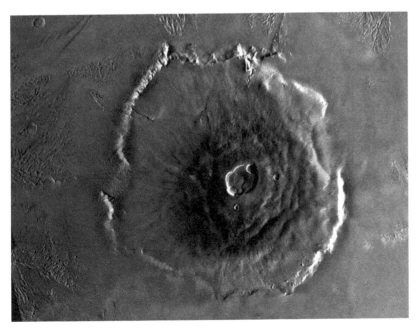

FIGURE 6.2 Olympus Mons, a large volcano on Mars as seen by the Viking Orbiter. The volcanic mound terminates at its base in a scarp or sharp cliff.
NASA/NSSDC

FIGURE 6.3 The large
volcanic crater (or caldera)
at the summit of Olympus
Mons shown in Figure 6.2.
    ESA/DLR/FU Berlin
(G. Neukum)

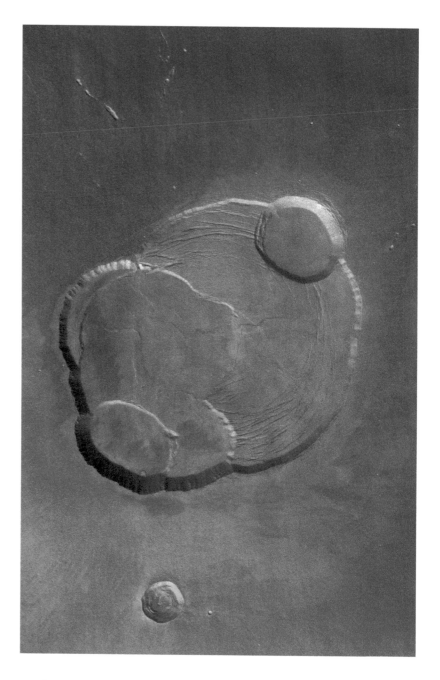

FIGURE 6.3 The large volcanic crater (or caldera) at the summit of Olympus Mons shown in Figure 6.2. ESA/DLR/FU Berlin (G. Neukum)

The roughly conical shape of the mountain and the presence of a crater at its summit show that it is a volcano. The summit crater, called a caldera, is a familiar feature of terrestrial volcanoes. It is formed by the collapse of surrounding material into the chamber of lava at the top of the volcano (Figure 6.3).

In every respect, the Mars mountain resembles the mounds of congealed lava that form on Earth when many successive outpourings of molten rock occur at a single spot over a period of millions of years. If the Pacific Ocean basin could be emptied, the Hawaiian Islands would be revealed as similar but smaller mounds of lava, rising out of the floor of the ocean basin in the same way that the Mars mountain rises out of the surrounding terrain.

About a dozen large volcanic mountains have been discovered on Mars (Figure 6.4). But there is evidence that the volcanoes have been extinct for a long time. The photographs show many modest-sized meteorite impact craters on the flanks of the volcanoes. These relatively small craters would have been obliterated if fresh floods of lava had run down the sides of the volcanic mountains in recent times. From the number of craters present, it has been estimated that the great Martian volcanoes have not been active for at least a few hundred million years.

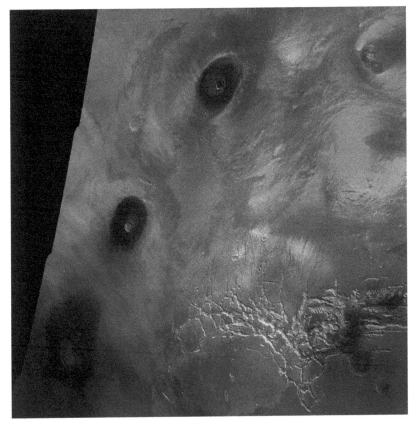

FIGURE 6.4 A view of Mars from the Viking 1 Orbiter showing three of the largest volcanoes (each about 200 miles or 300 kilometers in diameter) on the planet's surface.

NASA/NSSDC

FIGURE 6.5 The 3000 mile (4800 kilometer) long Valles Marineris canyon system (Viking Orbiter photograph). USGS

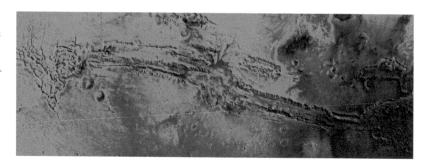

On Earth, volcanism is associated with the movement of large slabs of rock, generally thousands of miles in extent, that slide over underlying layers of warm and yielding material. These slabs of rock are called "plates." When the edge of one plate slides past another, cracks such as the San Andreas fault in California appear in the surface of the planet. When two plates collide head-on, their edges crumple, and huge masses of rock are thrust upward to form mountain ranges such as the Himalayas. When two plates move apart, a shallow cleft, called a rift valley, appears between them. The great Rift Valley, which runs through the Red Sea and continues into East Africa, is an example.

Are there also signs of the movement of plates on Mars, to accompany the evidence of Martian volcanism? Thus far, no clear evidence has been found of features like the San Andreas fault or the Himalayan mountains, but a feature that does resemble a long, straight crack or fault has been found. This is the Valles Marineris, a broad canyon 3000 miles (4800 kilometers) long, roughly 75 miles (120 kilometers) across, and 15 000 feet (5000 meters) deep at its lowest points (Figure 6.5).

Jumbled terraces of rock occur near the rim of this valley, which seem to be giant landslides formed by the slumping of material from the canyon walls (Figure 6.6). Their presence provides a clue to the process by which the Valles Marineris may have been formed. At some point in the geological evolution of Mars, forces deep within the planet must have pulled the crust apart along a line roughly coinciding with the present valley. Numerous small, parallel cracks appeared in the crust as a result of these forces, forming the beginning of the Valles Marineris. Repeated

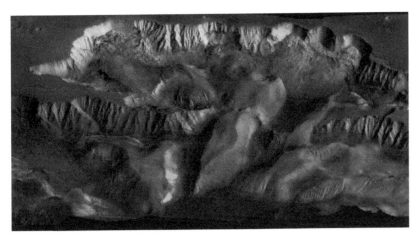

FIGURE 6.6 A portion of the Valles Marineris showing scalloped areas where sections of the canyon wall collapsed to form hummocky landslide deposits (Viking Orbiter mosaic).

NASA

slumping and landslides at the edges of the cracks gradually widened the Valles Marineris.

A similar process is believed to occur on Earth, when two plates in the lithosphere begin to break up and move apart. The separation of the plates containing the South American and African continents is the clearest example. Since there is, however, no clear sign of major plate movements on Mars, it is assumed that Martian tectonic activity never progressed beyond the incipient stage.

The absence of crustal movements on Mars provides a likely explanation for the large size of such Martian volcanoes as Olympus Mons. On Earth, when a pile of congealed lava accumulates on the surface to form a volcanic island, as in the Hawaiian Islands, the accumulation of lava never grows to a very great height because the movement of Earth's plates – in the case of Hawaii, the Pacific plate – carries the volcano away from the location of the rising lava. It is because of this circumstance that an entire chain of islands has developed in Hawaii; each one represents the accumulation of lava at the surface between these intermittent motions of the plates. If, however, there is no plate movement, the lava continues to pile up at the same place throughout geological periods of time, growing eventually to very great heights and covering a very large area.

The full range of elevation on Mars is about 19 miles (30 kilometers), about one and a half times the relief found on Earth. One of the most striking features of the global geology

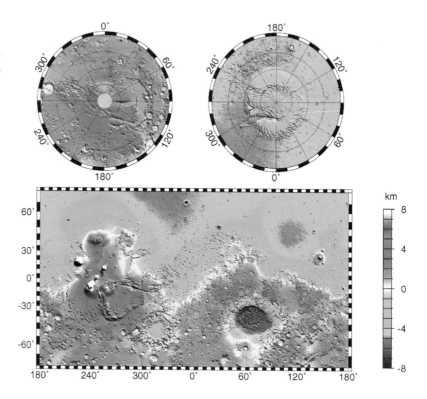

FIGURE 6.7 This high-resolution topographic map of Mars was generated by the Mars Orbiter Laser Altimeter (MOLA), an instrument aboard NASA's Mars Global Surveyor. The map clearly shows the relatively smooth lowlands of the northern hemisphere and the heavily cratered southern hemisphere. On the left, the great volcanoes of the Tharsis region and the Valles Marineris stand out. Large impact basins such as Hellas Planitia (right) and Argyre Planitia (left) are also visible.

MOLA Science Team

of Mars is the difference between the relatively smooth and sparsely cratered northern lowlands and the heavily cratered southern hemisphere, which is on average about 3 miles (5 kilometers) higher than the north (Figure 6.7). The northern hemisphere lowlands were most likely created by volcanic processes and may be the floor of an ancient ocean.

A number of large impact basins occur on Mars, including the massive Hellas Basin in the southern hemisphere, which is nearly 6 miles (9 kilometers) deep and 1300 miles (2100 kilometers) across. Hellas is surrounded by a rampart that rises 1¼ miles (about 2 kilometers) above the surroundings and extends out to 2500 miles (4000 kilometers) from the basin center (Figure 6.7).

6.1.3 Evidence for an early abundance of water

The orbit of Mars is about one and a half times the Earth–Sun distance. Surface temperatures get as low as −90 °C near the poles to −25 °C at the Martian equator. The combination of cold temperatures and low atmospheric pressure makes the

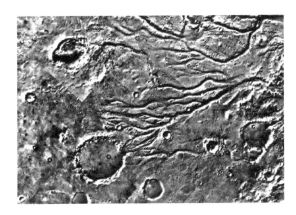

FIGURE 6.8 Viking Orbiter photograph of converging channels on the surface of Mars, resembling channels formed by water erosion in desert regions on Earth.
NASA

occurrence of liquid water impossible on Mars today. Any liquid water would simply evaporate immediately under the low pressure and then freeze out as ice. There is much evidence, however, that liquid water was present in abundance on the surface of Mars in ancient times.

Pictures taken by NASA spacecraft show converging channels resembling the channels cut by large volumes of water when flash floods occur in desert regions on Earth (Figure 6.8). Other indications of an early abundance of water include a braided pattern of channels, with streamlined islands, difficult to explain in any way other than by the flow of large volumes of water (Figure 6.9). Long, winding channels resembling riverbeds, with tributaries feeding into the main channel, can also be seen (Figure 6.10).

Recent spacecraft missions have discovered many areas that show evidence of layers of sediment. This implies a thick sequence of sedimentary layers indicating long periods of erosion and deposition by wind and water (Figure 6.11).

Photographs of the Martian channels thus provide evidence for the existence of a large amount of water on Mars in the past. But where is that water today? Direct evidence comes from measurements of the temperature of the Martian north polar cap and the humidity of the atmosphere above the cap, made from an orbiting spacecraft when the northern hemisphere was in its summer season (Figure 6.12). Prior to the first spacecraft measurements, the Martian polar caps were considered to be mainly composed of dry ice or frozen carbon dioxide, with a smaller admixture of ordinary ice or frozen water. However, under Martian atmospheric conditions the

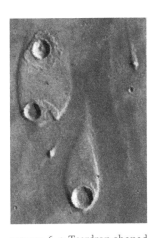

FIGURE 6.9 Teardrop-shaped islands, each about 25 miles (40 kilometers) long, rise above the floor of an ancient Martian outflow channel in this Viking Orbiter photograph. A torrent of water (which flowed from the bottom of this image toward the top) carried away substantial amounts of surface material but was deflected around the walls of the large impact craters shown here. The presence of small craters in the channel materials implies that billions of years have passed since the flow occurred.
ESA

FIGURE 6.10 A Martian feature about 200 miles (320 kilometers) long, resembling a dried riverbed with tributaries feeding into the main channel and a characteristic pattern of meanders. From the angle at which the tributaries enter the main channel, it can be concluded that the flow was from left to right in this Viking Orbiter photograph.
  NASA

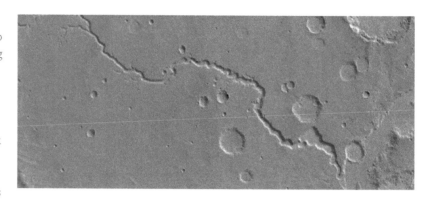

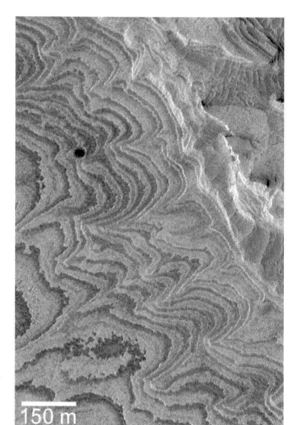

FIGURE 6.11 Layered sedimentary rock outcrops in Becquerel Crater of western Arabia Terra. These materials were deposited in the crater some time in the distant past, and later eroded to their present form. The layers could have been deposited directly from dust settling out of the Martian atmosphere or from water as the floor of an ancient lake.
  NASA/JPL/Malin Space Science Systems

temperature of a cake of dry ice would be about −100 °C. The measurements revealed the temperature of the northern polar cap to be −45 °C, which is too warm for dry ice but consistent with the interpretation that the caps are composed of frozen water.

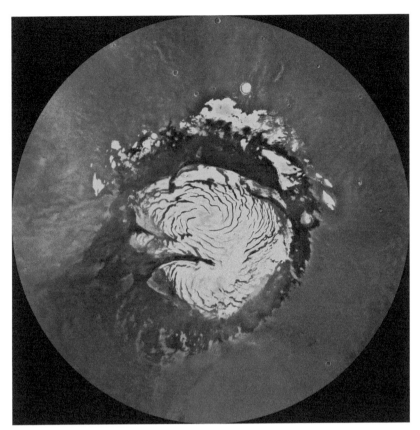

FIGURE 6.12 The north polar ice cap of Mars. During the summer, the carbon dioxide frost that covers northern latitudes during the winter evaporates, exposing a residual polar cap made primarily of water ice. The reddish streaks consist of dust, and the dark bluish areas are sand dunes that form a huge "collar" encircling the north polar ice cap (Viking 1 Orbiter image). NASA/USGS

The evidence firmly establishes that some of the water released to the atmosphere during earlier volcanic eruptions is still present on Mars, but locked up at the poles in frozen form. How much water is present? Estimates based on the volume of the polar caps give an upper limit of about 1 million cubic miles (4 million cubic kilometers), or about one and a half times the volume of the Greenland Ice Sheet. If this huge amount of ice were melted and spread over the surface of Mars, it would cover the planet with a shallow sea almost 100 feet (33 meters) deep.

A large amount of ice in the subsurface of Mars is also indicated by the presence of impact craters surrounded by masses of what appears to be a slurry of muddy soil ejected from the crater. These craters suggest that a shallow layer of ice underlies much of the planet's surface. In 2002, the Mars Odyssey spacecraft, in orbit around Mars, used neutron and gamma-ray spectrometers that can detect hydrogen (presumably in the form of H_2O) in the subsurface. These sensitive detectors discovered evidence for a widespread layer of ice less than 1 meter (3 feet) below the Martian surface.

These deposits were emplaced from the poles down to about 30° latitude in the northern and southern hemispheres of Mars. It is believed that the deposits were laid down in response to variations in the formation of ice and dust during ice ages produced by variations in the orbit of Mars. Some scientists think that these layered deposits are only about 2.1 to 0.4 million years old.

In some instances, these buried ice deposits seem to have melted suddenly, creating massive collapse features and torrents of water. It is possible that some of the Martian channels were carved by flash floods from these collapsed regions. The pattern of channels can be followed for hundreds of kilometers and the signs of the great flood peter out near the site where the first Viking spacecraft landed.

One of the most exciting discoveries of the Mars Odyssey spacecraft is the presence of youthful looking gullies on the walls of some impact craters (Figure 6.13). The gullies were apparently produced by erosion by liquid water escaping from a narrow layer in the shallow Martian subsurface. The freshness of the gullies suggests that they formed in relatively recent times.

From this evidence, a new and more complete picture of Mars emerges. Mars is a planet that experienced an early period of relative warmth and moisture, during which conditions may have been more congenial for the chemical evolution of life. Combining our knowledge of the present scarcity of water on Mars with the evidence for its abundance in the past, we are led to a tentative picture of the history of water on the planet. At one time, Mars may have had a substantial amount of water, which accumulated during a relatively long period of volcanism. During the period, its surface may have been traversed by streams and rivers that emptied into lakes or even small oceans. But, when the volcanism subsided, the source of the water disappeared. Over time, atmospheric gases and molecules of water leaked

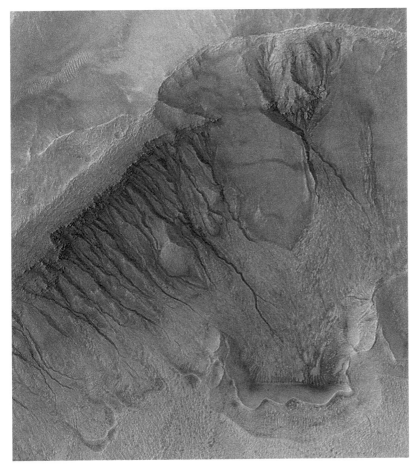

FIGURE 6.13 Gullies on the walls of an impact crater in Sirenum Terra, Mars (Mars Odyssey spacecraft image). The gullies in this crater originate at a narrow layer in the crater wall, and may have formed by the release of groundwater in geologically recent times.

NASA/JPL/MSSS

steadily away to space. Finally, the planet reached the condition in which we find it today, with a thin atmosphere with only a trace of moisture. A large amount of ice is present in the subsurface, but no liquid water flows on the surface of Mars.

6.2 Prospects for life on Mars

Does life exist on Mars? Mars is dry, cold, and less favorable than Earth for the support of life, but not implacably hostile. Experiments have shown that some terrestrial plants could exist in the Martian climate, although they would not flourish there. Because of the thin atmosphere, cosmic-ray and ultraviolet radiation bombard Mars with an intensity that would be lethal to terrestrial life. But biologists have suggested ways in which Martian organisms could have evolved natural shields against these deadly rays.

Water is the key element in a discussion of the prospects for Martian life. Water is an essential ingredient for the development of the kinds of life with which we are familiar, because it provides a fluid medium in which the complex molecules of the cell can move freely. This movement leads to frequent collisions between neighboring molecules and, as a consequence of these collisions, to chemical reactions that make up the ongoing process of life. The basic building blocks of living matter – immersed in the shallow Martian seas – would have collided ceaselessly; now and then the collisions would have linked them into the large molecules – proteins, DNA, and RNA – which are the essence of the living organisms. The linking of small molecules to form large ones would have marked the first step along the path from non-life to life.

The relationship between water and life lends a special interest to the spacecraft evidence for large amounts of water on Mars at an earlier time. If life exists on Mars today, it probably can be traced back to a golden age on the planet, when emission of gases from volcanoes maintained a substantial amount of liquid water on the surface, as well as a denser atmosphere. The transition to the drier climate of today may have occurred very slowly, over a period of millions of years and a very large number of generations. In this case, Martian life could have adapted progressively to the gradual onset of severe conditions. Life could exist on Mars today as a result of this long-continued process of natural selection if the planet once had an abundance of liquid water.

6.2.1 The search for Martian life

In July 1976, two Viking spacecraft landed on the surface of Mars, equipped with a variety of instruments intended for the study of Martian geology and biology. Each spacecraft contained a pair of television cameras providing stereoscopic

FIGURE 6.14 A Martian landscape photographed by the Viking 1 Lander. The horizon is about 2 miles (3 kilometers) from the spacecraft. Part of the upraised rim of a crater on the horizon is visible at the upper left.

NASA/JPL-Caltech

photographs of the nearby terrain. The cameras recorded a scene reminiscent of some of the desert areas of Earth, with rust-colored soil, sand dunes and other windblown formations, and scattered blocks of volcanic debris (Figure 6.14). Each spacecraft also contained a small, automated biochemical laboratory designed to carry out a series of tests for Martian life. The laboratory was operated by remote control from Earth and included instruments for the indirect detection of simple forms of life as well as the molecular building blocks of living matter.

All the tests used samples of Martian soil that had been scraped from the ground near the spacecraft by a small shovel mounted on the end of a boom (Figure 6.15). The samples were deposited in hoppers leading to separate chambers. The entire landing craft had been sterilized before the flight to ensure that any organisms the instruments detected would be indigenous to Mars, and not carried there from Earth.

The first test was based on the reasoning that all organisms known on Earth give off gases as waste products. For example,

FIGURE 6.15 The Viking experiments. The mechanical arm with its small scoop protrudes from the right side of this view of the Viking 1 Landing site. Several small trenches dug by the scoop in the Martian soil appear near the left side of the picture.

NASA

plants release oxygen as a waste product, and animals and most kinds of microbes release carbon dioxide. If the Martian soil contained organisms resembling plants, a pinch of soil placed in a chamber would produce a small amount of oxygen. If the soil contained microbes or animals, carbon dioxide would be produced. An instrument adjacent to the chamber could detect this and other gases.

When the test for gases was carried out, carbon dioxide was released from the soil, indicating that microbes might live in it. At the same time, a substantial amount of oxygen appeared, suggesting that the soil might contain plant life. The carbon dioxide was released slowly and steadily over a period of several days, as would be expected from a population of microbes in the soil. The oxygen, however, came off in a burst in the first few hours of the experiment. This was a surprise, because if plants were producing the oxygen, it would be released at a slow, steady rate, like the carbon dioxide.

The very rapid release of oxygen was more consistent with a chemical reaction in the soil, rather than a biological process. One kind of chemical reaction that could release a burst of oxygen is known from laboratory experiments. Solar ultraviolet radiation – falling on a surface of soil – can produce molecules of hydrogen peroxide, or H_2O_2, which then adhere to the grains of rock in the soil. If the grains of rock are moistened, the hydrogen peroxide molecules break up very quickly into water and oxygen. Since the initial step in the Viking experiment was the exposure of the soil to moisture, the release of oxygen could be explained readily by the chemical process just described, without recourse to life processes.

Another test was designed primarily for microbes. In this test, a pinch of soil was placed in the chamber and moistened with a nutrient broth containing amino acids and other food substances. These substances, like all foods, were made from atoms of carbon, oxygen, nitrogen, and other chemical elements. If microbes existed in the soil, they would consume the food and digest it. Most of the atoms in the food would be incorporated into the bodies of the microbes as they grew and reproduced, but some carbon atoms would be released into the atmosphere in the form of molecules of carbon dioxide as a by-product of their metabolism.

How could scientists on Earth find out whether that complicated process was taking place in a chamber on Mars, more than 200 million miles (320 million kilometers) away? The answer is ingenious. Before the flight to Mars, a special kind of food had been prepared, in which some of the carbon atoms were the radioactive isotope of carbon, carbon-14. If Martian microbes were present, they would digest the carbon-14 atoms and exhale radioactive carbon dioxide. To find out if this were happening, a detector sensitive to radioactivity was placed in an adjoining chamber, located above the first one and separated from it by a thin tube. If the detector signaled the arrival of radioactive carbon dioxide, that would suggest that Martian microbes existed in the chamber below.

When the test was carried out, the detector indicated that a large amount of radioactive carbon dioxide had been produced in the pinch of soil below, and had passed through the tube into the upper chamber. Thus, this test suggested the presence of Martian life.

But the test for microbes, like the first test, also could be explained by a chemical reaction not involving organisms. Suppose Martian soil contains hydrogen peroxide, as seemed to be indicated by the release of a burst of oxygen in the previous experiment. When the soil was moistened by the nutrient broth, the peroxide compounds would decompose the particles of food in the broth, breaking them up into smaller molecules, including molecules of carbon dioxide. The carbon dioxide would be released to the atmosphere. Some of this carbon dioxide would be radioactive. In that way, chemical reactions could simulate the presence of microbes in the second test. However, repetitions of the microbe test under different conditions yielded results that are difficult to explain by chemical reactions, unless the reactions are quite complicated.

Additional information came from still another experiment, not designed specifically as a test for life, which tested the soil for complex compounds containing carbon. These compounds, known as organic molecules, are the building blocks of all life on Earth. If life exists on Mars and is similar

FIGURE 6.16 This view from the Mars Pathfinder landing site shows the rock field surrounding the spacecraft. In the foreground is one of the three solar panels used to power the spacecraft, along with the yellow ramp along which the Sojourner rover rolled down onto the surface. The rover, the size of a child's toy wagon, can be seen next to a rock nicknamed Moe.
NASA/NSSDC

to terrestrial life, there should be an abundance of organic molecules in the Martian soil. But the test failed to reveal any organic molecules within the limits of sensitivity of the instrument. That result suggested that life does *not* exist on Mars, and tended to support the chemical explanation for the Viking results. The final resolution of the question of Martian life may have to await more elaborate tests, in future landings on the planet's surface.

6.2.2 Pathfinder mission to Mars

The Pathfinder mission in 1996 represented a return to the surface of Mars after 20 years, but the joint package of a lander and rover (named Sojourner) did not include life-detection experiments (Figure 6.16). The rover was instead equipped with a camera and an X-ray spectrometer designed to measure the chemical composition of the rocks lying on the surface. Since these rocks may have been dispersed by ancient flood waters, scientists hoped that they would represent a good sampling of the Martian surface rocks. Sojourner found the rocks to be mainly basalts, but some rocks contained an unusually high percentage of silicon. On Earth, such rocks would represent the end product of a considerable amount of melting and re-melting, so these rocks could indicate a long history of geological activity for the planet.

6.2.3 The Mars exploration rovers

In early 2004, two spacecraft again landed on Mars. The Mars Exploration Rover mission put two landers – Spirit and Opportunity – on opposite sides of Mars in areas that appear to have been affected by liquid water in the past. The landers released rovers equipped with instruments to determine the chemical composition and mineralogy of the rocks and soil. Each rover is designed to operate like a robot geologist walking the surface of Mars, examining and mapping the areas around the spacecraft.

Spirit landed at Gusev Crater, on a possible former lakebed within an impact crater (Figure 6.17). Some scientists think that Gusev Crater was once filled by a body of water in which thick sediments were deposited. Opportunity touched down in Meridiani Planum, a flat plain where deposits of the iron-rich mineral hematite (Fe_2O_3) hinted at the former presence of liquid water (Figure 6.18).

6.2.4 Ancient life in a Martian meteorite?

Although no Mars rocks have ever been returned to Earth by spacecraft, two dozen meteorites are known that originated on Mars. The clue to their origin comes from inclusions of gases that are identical to those in the Martian atmosphere. They also have ages that are much younger than meteorites that come from the asteroid belt or the Moon. These Martian

FIGURE 6.17 Spirit Lander on Mars in Gusev Crater, an area thought to be an ancient lakebed. The round, dark markings in the central flat area may have been left by the lander as it bounced across the surface.

NASA

FIGURE 6.18 Opportunity Lander at Meridiani Planum. The dark undulating terrain is believed to be rich in hematite (Fe_2O_3), an iron-bearing mineral that can form in watery environments. Marks in the foreground were produced by the spacecraft.

Mars Exploration Rover Mission, JPL, NASA

meteorites are chunks of the volcanic crust of Mars that were ejected from the planet's surface by large impact events. One such meteorite – labeled ALH 84001 – was discovered in the Allen Hills ice field in Antarctica in 1984. It is a piece of Mars that formed 4.5 billion years ago, was fractured by an impact about 3.8 billion years ago, and ejected from Mars by another impact about 16 million years ago (Figure 6.19). After orbiting the Sun for millions of years, ALH 84001 fell in Antarctica about 13 000 years ago.

In 1996, NASA scientists studying ALH 84001 reported that the meteorite contained evidence for ancient Martian life. Their analyses detected grains of carbonate, possibly formed in liquid water, pure crystals of magnetite (iron oxide) such as are produced by terrestrial bacteria, organic carbon compounds, and tiny structures that resemble simple micro-organisms (Figure 6.20). However, the interpretation of these findings is controversial, and many scientists think that the materials found in the meteorite were produced by strictly non-biological processes.

6.2.5 The significance of the search for Martian life

The discovery of Martian life – even in as simple a form as a microbe – would have portentous implications for the probability of life on other earthlike planets in the Cosmos. At the present time, no one knows what that probability is.

FIGURE 6.19 Meteorite ALH 84001 is thought to have been formed on Mars 4.5 billion years ago, dislodged from Mars by a massive impact about 16 million years ago, and landed on Earth 13 000 years ago. The small cube with letters on it is 1 centimeter (0.4 inch) across.
   NASA

FIGURE 6.20 This image, magnified some 100 000 times, shows elongate structures found within the Martian meteorite ALH 84001. Some scientists have interpreted these as fossils of micro-organisms that lived on Mars billions of years ago. This interpretation is controversial, however, and the structures may be purely inorganic.

David S. McKay, NASA

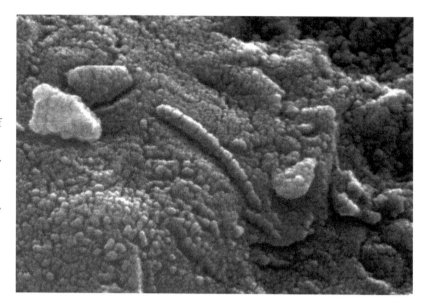

Perhaps the evolution of life out of inanimate atoms requires the simultaneous occurrence of many special circumstances, whose combined probability is so low that this event has happened only once in the Universe. How can we determine whether this is so?

The discovery of life on a second planet in our Solar System would go far toward settling that question. Life on *one* planet – Earth – tells us nothing about the probability of life in the Universe, but life on *two* planets in one solar system – Earth and Mars – would tell us nearly everything. For, if life has arisen independently on two planets in a single solar system, it cannot be a rare accident, but must be a fairly probable event. If life is found on Mars, that will suggest that countless inhabited planets may surround us in the Milky Way Galaxy and other galaxies. No scientific discovery more significant in its astrobiological implications can be imagined.

6.3 Summary

By the end of this chapter you should be familiar with the following concepts and topics:

The possibility for life on Mars, either now or in the past
Detailed understanding of the planet
  Geology of Mars
  Evidence for geological activity
  Evidence for an abundance of liquid water and substantial
    atmosphere
  Evidence for frozen water beneath the surface of the planet
    today
Mars as a focus for astrobiology and the search for life
  The search for Martian life
  The Mars Exploration Rovers
  Ancient life in a Martian meteorite?
  The significance of the search for Martian life

Questions

6.1. Explain the following statement: Mars may be experi-
     encing today the geological conditions that the Earth will
     experience several billion years from now.
6.2. What are the present surface conditions on Mars?
6.3. What were the results of the search for Martian life?
6.4. What would be the significance of finding life on Mars?

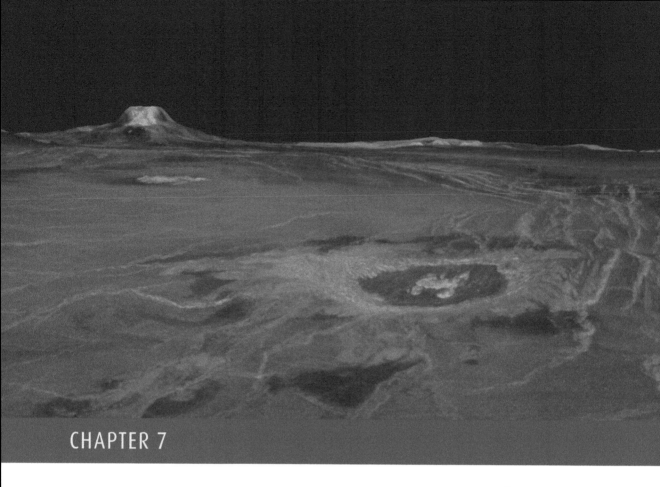

CHAPTER 7

Venus – our sister planet

The evolution of a hostile world

Ever since Galileo first turned his telescope on Venus in 1610, astronomers have wondered whether there may be life beneath its dense covering of clouds. Venus is Earth's sister planet in the sense that it is similar in size and mass, and situated only slightly closer to the Sun. But recent mapping of the surface has shown that the geology of Venus is rather different from that of Earth, and although it does have an atmosphere, it is hugely different from our own. Why are conditions on Venus so different from those on Earth, when the starting points were so similar? Is there any potential for life to exist there?

In this chapter we look at results from spacecraft missions to Venus, which have documented its geology and the conditions beneath the thick covering of cloud. The surface of Venus is dotted with volcanoes, and we will learn how scientists have come to the conclusion that there has been recent volcanic activity on the planet. We will discover how Venus has evolved into a hostile world so different from Earth, with an excessively high temperature and pressure that make the existence of any form of life highly unlikely. The chapter ends with a discussion of the habitable zone and the factors that determine the habitability of a planet.

7.1 Venus

The densities of Venus and Earth are very similar, indicating that they probably have a similar bulk composition, dominated by iron and rocky materials. As a consequence, Venus can be assumed to have a similar concentration of radioactive elements. Because Venus is close to Earth in size, it should seal in its internal heat about as well as our planet, and should have a similar geology, with a partly molten zone at some depth below the surface, a solid crust, and plate movements. There should also be zones of earthquakes, volcanoes, and mid-ocean ridges.

However, it has turned out that the geology of Venus is rather different from that of Earth. Mapping of the planet's surface has found no evidence for the kind of crustal movements that are known to generate the large-scale features of Earth's surface.

The first map of the surface of Venus was compiled from elevation data from radar signals bounced off the planet by the Pioneer-Venus spacecraft in the late 1970s. This map revealed two large areas that stand out above the surrounding terrain. One, called Ishtar Terra, is comparable in size to the United States. Ishtar Terra contains a mountain massif called Maxwell Montes, which is higher than Mount Everest. Its summit is 35 400 feet (about 12 000 meters) above "sea level" on Venus – that is, above the average elevation of the Venus surface. The second highland region, called Aphrodite Terra, sits on the equator, and is about the size of North Africa (Figure 7.1).

Previous page: A portion of western Eistla Regio is displayed in this three-dimensional perspective view of the surface of Venus, produced at the JPL Multimission Image Processing Laboratory.
    NASA

FIGURE 7.1 Elevation map of Venus compiled from Pioneer-Venus radar altimeter data. Colored contours indicate surface elevations on the planet. Lowland regions (dark blue) and highlands (reds, yellows, and light greens) cover about one-third of the planet's surface. The remainder (light blues and dark greens) is rolling upland terrain that has close to the average level of Venus' surface. Large elevated regions are Ishtar Terra (in the northern hemisphere) and Aphrodite Terra (in the equatorial latitudes)

NASA-NSSDC

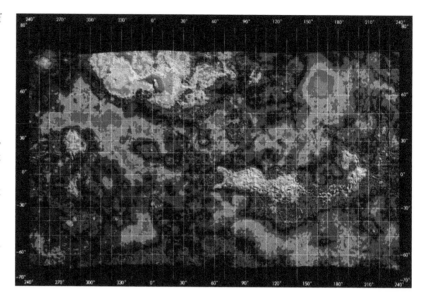

The more detailed radar images from the Magellan spacecraft that orbited Venus in the early 1990s revealed that the face of Venus is dominated by extensive plains covered by volcanic lava flows (Figure 7.2). Large rift valleys wind over the surface and converge on broad high rises (Figure 7.3). Volcanoes, like those of Earth and Mars dot the surface.

Several areas on the planet show signs of recent volcanic activity, and may still be actively erupting. These volcanoes probably sit on top of a localized volcanic hotspot, similar to the volcanic hotspot that lies under the island of Hawaii on Earth (Figure 7.4).

Circular dark features seen in the radar images are large meteorite impact craters (Figure 7.5). The Magellan spacecraft revealed more than a thousand impact craters ranging up to 180 miles (290 kilometers) in diameter. Few small craters exist because small asteroids and comets burn up in the thick atmosphere of the planet. Craters on Venus are spatially random, indicating that all large areas of the planet are about the same age. From the numbers of impact craters present, scientists have been able to determine that most of the surface on Venus was covered by lava flows within the past 600 to 700 million years. This relatively young volcanism supports the idea that Venus is still geologically

FIGURE 7.2 In this Magellan radar image of Venus, Aphrodite Terra is the elongated, light-colored, wispy feature that wraps one-third of the way around the planet. Aphrodite Terra is roughly parallel to Venus' equator. In this and other Magellan images, simulator colors are based on color images recorded by the Russian Venera 13 and 14 spacecraft.

Magellan Project, JPL, NASA

active. More direct evidence for active volcanism on Venus comes from Magellan spacecraft observations of sudden changes in the amount of sulfur in the Venus atmosphere. Sulfur dioxide is released in large quantities when volcanoes erupt explosively.

7.2 Surface conditions

Some properties of Venus suggest that it should have provided an agreeable climate for living organisms. The planet is 67 million miles (108 million kilometers) from the Sun, while the distance of Earth is 93 million miles (150 million kilometers). Because Venus is closer it receives approximately twice the intensity of sunlight reaching Earth. However, the heavy cloud cover on Venus keeps out 80% of this excess of

FIGURE 7.3 Magellan radar image of Beta Regio, Venus. Beta Regio consists of two large volcanic structures: Theia Mons, the bright nearly circular feature to the south (north is up), and Rhea Mons, the bright feature to the north. The distance between them is about 500 miles (800 kilometers).

NASA-NSSDC

solar energy (Figure 7.6). As a consequence, the average temperature on the planet should be only slightly higher than the average temperature on Earth, and very comfortable by terrestrial standards.

As early as 1956, however, radio astronomers obtained evidence which suggested that the climate on Venus might be far from balmy. In fact, the measurements indicated that Venus is a very hot planet – with temperatures reading a sizzling 500 °C.

This temperature is hot enough to melt lead. It is also high enough to break apart all the delicate molecules that make up the essential ingredients of a living cell. Thus, it is unlikely that organisms remotely resembling any form of life on Earth could exist on the surface of Venus.

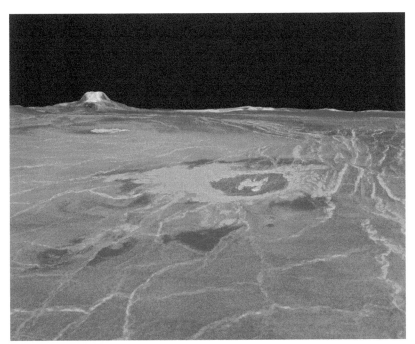

FIGURE 7.4 A portion of the Eistla Regione area of Venus imaged by the Magellan spacecraft radar showing the 9000-foot (3000-meter) high volcano Gula Mons in the background. In the foreground is the 30-mile (50-kilometer) diameter impact crater Cunitz.
NASA, JPL

FIGURE 7.5 Magellan spacecraft radar image of three impact craters in the Lavinia Planum region of Venus. The crater in the foreground has a diameter of 23 miles (37 kilometers).
NASA, JPL

FIGURE 7.6 The cloud-
covered face of Venus as seen
by the Mariner 10 spacecraft.
This photograph, taken at
ultraviolet wavelengths,
shows variations in cloud
structure produced by the
rapid circulation of the
atmosphere.

   NASA

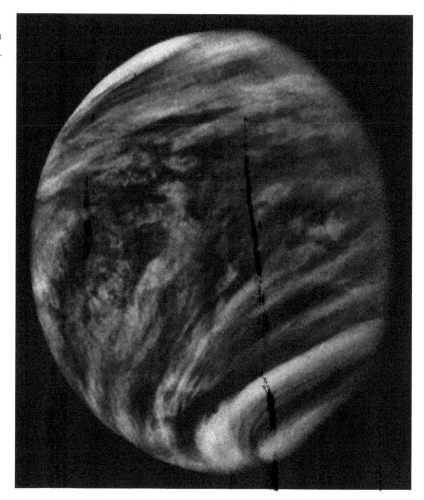

7.3 The Venus atmosphere

When spacecraft reached Venus, direct temperature measurements showed that the temperature of the surface is, in fact, extremely high with an average of about 600 °C. The measurements also indicated that the atmospheric pressure at the surface is nearly one ton per square inch (92 bar) – almost 100 times the pressure on the surface of Earth.

Finally, the observations revealed that carbon dioxide makes up about 96% of the atmosphere, compared to 0.03% in Earth's atmosphere. Nitrogen, which constitutes 78% of Earth's atmosphere, is only 2 to 4% of the Venus atmosphere. With allowance for the greater density of the Venus atmosphere, the absolute amount of nitrogen is approximately the same on Venus and Earth. This result supports the theory that the atmosphere of an earthlike planet consists of gases from the planet's interior, which are released at the surface in volcanic eruptions. Another source of water and other volatiles is impacts by icy comets during the late stages of the planet's accretion. If Venus and Earth experienced the same amount of volcanic activity during their lifetimes, they should have similar amounts of nitrogen in their atmospheres. They should also have similar amounts of other volcanic gases, such as water vapor. The spacecraft observations, however, revealed the presence of only a small amount of water vapor, sufficient, if condensed to liquid form, to cover the surface of the planet to a depth of 1 foot (0.3 meters).

Spacecraft have recorded conditions beneath the clouds and down to the planet's surface. At an altitude of about 40 miles (64 kilometers), a smog layer of sulfur particles covers the planet. Below the smog lies a thick cloud layer composed of sulfuric acid droplets. These clouds of sulfuric acid droplets or *aerosols* are apparently produced by explosive volcanic eruptions on Venus, which put large amounts of sulfur-rich gases high into the atmosphere. The clouds thin out below 20 miles (32 kilometers) and the atmosphere becomes clear. The Sun's light grows weaker as the atmosphere approaches the density of soup. At the surface the intensity of sunlight corresponds to a heavily overcast day on Earth, and the light has a strong orange-brown cast.

7.3.1 The greenhouse effect

Why is the surface of Venus so hot? What controls the temperature on the surface of this planet, or any planet? A planet's temperature is determined primarily by its distance from the Sun, which controls the intensity of the solar radiation falling on the planet's surface. Most of this energy reaches the planet in the form of visible light. A part of the light is reflected back to space by clouds and scattered in the atmosphere; the remainder passes through the atmosphere to the surface, which is warmed by the absorption of this solar energy. The warm surface radiates energy back to space in the form of infrared radiation or heat. Over a long period of time, a planet must radiate as much heat back to space as it receives in the form of visible sunlight.

At first, the surface of the planet will rise steadily in temperature as it receives the solar energy, but as it becomes hotter it radiates more heat to space, until eventually the outgoing heat just balances the inflow of sunlight. For any planet, the temperature can be calculated at which the radiated heat just equals the influx of sunlight. In the case of Earth, we find in this way that the average ground temperature should be $-30\,°C$.

But the actual average temperature of Earth is $15\,°C$, and not $-30\,°C$. The increase in temperature is produced by our atmosphere, which acts as an insulating blanket, trapping a part of the radiation from the ground and returning it to the planet, where it adds to the heat provided by the absorption of sunlight. In this way, the average temperature of Earth is raised from the theoretical value of $-30\,°C$ to the level of $15\,°C$, which is actually observed.

The outgoing heat is trapped by the trace amounts of water vapor, carbon dioxide, and ozone in the atmosphere. Although these constituents together make up less than 1% of Earth's atmosphere, they absorb 90% of the heat radiated into the atmosphere from the surface.

The heat insulation provided by the atmosphere is called the *greenhouse effect*. A greenhouse has a glass cover that is transparent to the Sun's visible radiation (just as Earth's atmosphere is transparent), but blocks the heat radiated from the ground within. Thus the heat is trapped within the greenhouse and warms the interior, just as water vapor, carbon dioxide, and ozone retain heat in Earth's atmosphere and warm the surface of our planet (Figure 7.7).

The greenhouse effect provides the answer to the mystery of Venus' high temperature. Carbon dioxide, of which only a trace exists in Earth's atmosphere, is present in the atmosphere of Venus in such massive amounts – 100 000 times more than in Earth's atmosphere – that it makes the Venus atmosphere almost impervious to infrared radiation. Nearly all the heat radiated from the surface of Venus is trapped by

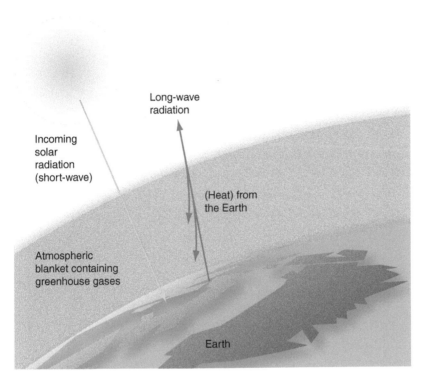

Incoming
solar
radiation
(short-wave)

Long-wave
radiation

(Heat) from
the Earth

Atmospheric
blanket containing
greenhouse gases

Earth

FIGURE 7.7 The greenhouse effect. Visible light from the Sun passes freely through the atmosphere to the surface of the planet. The surface, heated by the Sun's rays, sends out infrared radiation, which does not pass as freely through the atmosphere. Some is absorbed by greenhouse gases in the atmosphere, heating it, and the heated atmosphere, in turn, warms the planet.

the insulating atmospheric blanket composed of carbon dioxide, and smaller amounts of water vapor and sulfur dioxide. The remainder is trapped by clouds and haze. The trapped heat causes the temperature of the surface to rise far above the value it would have if the atmosphere of Venus were similar to ours.

The abundance of carbon dioxide on Venus explains its high temperature. However, this explanation leads to another question: Why does Venus have so much carbon dioxide? Why did Earth and Venus, formed out of similar materials, develop different atmospheres?

The answer emerges from an examination of the factors that govern the evolution of a planet's atmosphere. The present atmospheres of the terrestrial planets were released as volcanic vapors during the course of their geological evolution. Carbon dioxide is an abundant gas in volcanic eruptions on Earth, second only to water vapor. Since most of the water vapor condenses to form liquid water, the carbon dioxide, which remains in the atmosphere, should be the dominant atmospheric gas on Earth. In fact, Earth should be covered with a dense blanket of carbon dioxide, exerting a pressure of half a ton per square inch (50 bar) at the surface.

Assuming that Earth and Venus have had similar histories, that should also be true of the Venus atmosphere. The atmosphere of Venus is observed to have just the amount of carbon dioxide predicted by this reasoning. However, Earth has 100 000 times less. So now we see that the problem is not to explain why Venus has so much carbon dioxide in its atmosphere, but why Earth has so little.

7.3.2 Removal of carbon dioxide from Earth's atmosphere

The answer to the question of the missing carbon dioxide in Earth's atmosphere is connected with the fact that the rocks of Earth's crust are able to absorb carbon dioxide. The absorption takes place in a chemical reaction in which atmospheric carbon dioxide combines with rocks to form substances called *carbonates*, somewhat as oxygen in the atmosphere combines with iron to form iron oxide or rust. Geochemists have long been familiar with chemical reactions (named "Urey reactions" after the geochemist Harold Urey) taking place at the surface of Earth that demonstrate this process. An example of such reactions is

$$CO_2 + CaSiO_3 \rightarrow SiO_2 + CaCO_3$$

in which carbon dioxide reacts with calcium silicate, a mineral common in surface rocks, to form silicon dioxide plus calcium carbonate. For the reaction to proceed, however, liquid water must be present.

This and similar reactions, going on slowly but steadily at the surface of Earth over millions of years, presumably removed carbon dioxide from the atmosphere and converted it into solid compounds nearly as fast as it entered the atmosphere through volcanic eruptions.

A partial confirmation of this theory is found in the fact that the amount of carbon locked up in the sedimentary rocks in Earth's crust in the form of carbonates (i.e., the rock called limestone) is approximately equal to the amount of carbon contained on Venus in the form of atmospheric carbon dioxide.

Perhaps the removal of carbon dioxide by the rocks of Earth's crust was the reason why the temperature on Earth stayed at a comfortable level long enough for life to take hold here. As soon as life appeared in abundance, its own chemistry would have further reduced the level of carbon dioxide in Earth's atmosphere. Marine animals absorb carbon dioxide from seawater and convert it within their bodies to solid carbonates (seashells, for example, are nearly pure calcium carbonate). For every molecule of carbon dioxide removed from the water by this process, one molecule eventually enters the ocean from the atmosphere. Thus, the absorption of carbon dioxide from seawater by marine animals is equivalent to its removal from the atmosphere.

7.3.3 The runaway greenhouse effect

The rocks on the surface of Venus must contain the same silicate minerals that removed carbon dioxide from Earth's atmosphere. Why was carbon dioxide not absorbed on Venus?

The answer is that the reactions that remove carbon dioxide depend on the presence of liquid water, which is dependent on temperature. When Earth was a young planet, it was cool enough to have liquid water on the surface. This allowed the chemical reactions that remove atmospheric carbon dioxide to take place.

When Venus, on the other hand, was a young planet with only a thin atmosphere, its average temperature was higher – about $60\,°C$ – because of its closeness to the Sun. At this higher temperature, water vapor released by volcanism remained in the atmosphere and the rocks on its surface absorbed far less carbon dioxide than was absorbed by rocks on Earth in the same period. According to the measurements on the temperature dependence of the reactions, the amount of carbon dioxide remaining in the atmosphere of Venus must have been 10 to 100 times greater than the amount in Earth's atmosphere.

This blanket of carbon dioxide, although still relatively thin at that early time in comparison with the heavy carbon dioxide atmosphere on Venus today, would have been great enough to produce an appreciable greenhouse effect. The greenhouse effect, raising the temperature of the surface of Venus still higher, would further limit the availability of liquid water and further diminish the effectiveness of the Venus rocks in absorbing the carbon dioxide that continued to pour into the atmosphere through volcanoes. As more carbon dioxide accumulated in the atmosphere, the greenhouse effect became stronger, the temperature of the planet rose higher, the effectiveness of the reactions that absorb carbon dioxide was still further diminished, the greenhouse effect was further enhanced, and so on. In this way, a runaway greenhouse effect started, leading to the massive carbon dioxide atmosphere and ovenlike conditions that characterize Venus today.

7.3.4 Why oxygen is present in Earth's atmosphere but not on Venus

These ideas explain the difference in the amounts of carbon dioxide in the atmospheres of Earth and Venus. They do not account for the presence of free oxygen, the second most abundant constituent of our modern atmosphere. This gas should not exist in the atmosphere in appreciable concentrations despite its high abundance in the Cosmos, because its strong tendency to enter into chemical reactions removes it promptly from the atmosphere and deposits it on the surface in the form of solid oxides. The explanation lies in the presence of life on Earth. The main source of atmospheric oxygen is generally believed to be plant life. Plants absorb carbon dioxide from the atmosphere, convert the carbon into compounds that enter the structures of their bodies, and return the residual oxygen atoms to the atmosphere.

If plant materials are removed from Earth's surface by burial before the organic debris is re-oxidized by bacterial action, then free oxygen can accumulate in the atmosphere. These biological activities continually renew the supply of free oxygen in the atmosphere, which would otherwise be depleted rapidly – in 2000 years or so – by chemical reactions with the materials of Earth's crust.

7.3.5 Why Venus has less water than Earth

Water, like carbon dioxide, was one of the gases that must have been trapped in the interiors of both Earth and Venus when they first formed out of the solar nebula. The release of trapped water vapor through cracks and fissures in the crust of Earth has led to the accumulation of oceans covering three-quarters of the surface of the planet to a mean global thickness of 8000 feet (2700 meters).

A comparable amount of water should exist on the surface of Venus. Because of the planet's high temperature, this water should be in the form of vapor in the atmosphere, rather than a liquid filling the natural basins in the topography of the crust. But it is known from the results of spacecraft measurements that the equivalent of an ocean of water is not present on Venus. In place of the 8000 feet of liquid water that cover the face of Earth, the atmosphere of Venus has no more than the equivalent of a few feet of liquid water.

One package of instruments dropped into the Venus atmosphere by the Magellan spacecraft included a device for making accurate measurements of the masses of atoms and molecules in the atmosphere. The measurements yielded the proportion of heavy water molecules to ordinary water molecules in the Venus atmosphere. The heavy water molecule is one in which one or both of the hydrogen atoms are replaced by atoms of deuterium (heavy hydrogen). During the history of the atmosphere of a planet such as Earth or Venus, water is lost continually from the upper levels of the atmosphere because solar ultraviolet radiation breaks the water molecules into their basic hydrogen and oxygen atoms, and the hydrogen atoms, being relatively light, escape to space. The oxygen atoms remain behind, because they are heavier and are held by the planet's gravity, and enter into chemical combinations with elements in the crust.

As a result of this process, water is removed steadily from the surface and atmosphere of an earthlike planet. On Earth, however, temperatures in the middle atmosphere are so cold that water freezes into ice crystals and falls out of the atmosphere before reaching the zone of intense solar ultraviolet radiation.

When the atmosphere of a planet contains a mixture of ordinary water molecules and heavy water molecules, the ordinary water is lost to space more rapidly than heavy water, because the hydrogen it contains is a lighter atom and escapes more rapidly. Thus, as time goes by, both ordinary water and heavy water escape, but since the ordinary water escapes more rapidly, the proportion of heavy water to ordinary water gradually rises.

A measurement of the relative amounts of heavy water and ordinary water on Venus should enable us to infer how much water has escaped from the planet during its history. The measurements show that a great deal has escaped; in fact, the amount of water now present in the atmosphere of the planet is only a hundredth of what was originally present.

7.4 Conditions for life: The habitable zone around a star

How far from the Sun must a planet be to maintain a congenial climate for life? The exploration of Venus has told us that the planet was too close, and that was its undoing. If it had been nearly as far away as Earth is, the temperature of its surface might have climbed slowly enough to permit life to gain a toehold, and once life began, it could have held the abundance of carbon dioxide in check, and might have prevented the temperature from climbing out of bounds subsequently. But Venus, having lost its chance to harbor life when the Solar System was first formed, could never again recapture the opportunity.

The comparative study of the histories of the atmospheres of Earth and Venus demonstrates that an earthlike planet must not be too close to its sun if conditions are to remain favorable for the evolution of terrestrial forms of life. In the case of our Solar System, the critical threshold lies somewhere between the orbits of Venus and Earth. For our Sun, that is the inner edge of the habitable zone.

How far on the other side of Earth's orbit can a planet be, and yet maintain a comfortable climate for the support of life? The answer depends mainly on the size of the planet. It may be quite far from the Sun, and the warming effect of the Sun's rays correspondingly small, but if the planet is large enough to seal in its radioactive heat and build up a high level of internal temperatures, volcanic eruptions will be frequent, and copious emissions of gases can be expected from the planet's interior. Thus, its atmosphere will accumulate to a high density, and a significant greenhouse effect will develop. The greenhouse effect will trap the heat of the Sun and create a comfortable temperature on the planet's surface, in spite of its distance from the Sun and the weakness of the solar radiation.

If, however, the planet is relatively small – the size of the Moon, for example – its internal temperatures will not rise to a sufficiently high level to produce substantial volcanic activity with significant emission of gases. Moreover, the relatively weak gravitational field of the planet will render it less effective in preventing the lighter gases from escaping to space, further diminishing the density of the atmospheric blanket. For these two reasons, the greenhouse effect will be too small to increase the temperature of the planet's surface appreciably. The scarcity of water – as a further consequence of the diminished emission of gases from the interior of the planet – will also tend to act against the onset of biological evolution.

Mars, it seems, was a marginal case. Mars is one-tenth as massive as Earth, but ten times more massive than the Moon. Mars was large enough to remain geologically active for several billion years. As long as the volcanism persisted, the planet probably possessed a relatively comfortable environment for life. But eventually its internal heat dissipated, and Mars grew cold and dry. An interesting conclusion follows: Mars is a less congenial place for life than Earth, not primarily because it is farther from the Sun, but because it is smaller and geologically less active. If a giant hand moved Earth out to the orbit of Mars, the average temperature of our planet would drop. However, Earth's insulating blanket of air, continually replenished for billions of years, would maintain comfortable temperatures over large areas. The chill would be noticeable, but life could continue. Thus, the outer edge of the habitable zone depends on the size and geological activity of an earth-like planet.

7.5 Summary

By the end of this chapter you should be familiar with the following concepts and topics:

Venus
Spacecraft missions to Venus
  Documentation of its geology
  The conditions beneath the thick covering of cloud
  Surficial volcanoes
  Recent volcanic activity on the planet
Evolution of Venus into a hostile world
  Excessively high temperature
  Excessively high pressure
  The greenhouse effect
  Why oxygen is present in Earth's atmosphere but not on
    Venus
  Why Venus has less water than Earth
The habitable zone and the factors that determine the
    habitability of a planet

Questions

7.1. What are the surface conditions on Venus today?
7.2. Why are the temperatures so high in the Venus atmosphere?
7.3. Explain the term "habitable zone."

PART III

The Earth

CHAPTER 8

Earth

Composition and structure of a habitable planet

A chain of cause and effect connects Earth and all its inhabitants to events stretching back over billions of years to the early days of our Universe. We can understand what Earth and we ourselves are made of in terms of the life histories of all the stars that lived and died in the Universe before the Solar System and Earth were formed. Our planet is composed of some of the most common elements in the Universe, which were produced in stars and spread into space by supernova explosions. As Earth is largely made up of such common elements, perhaps earthlike planets with the potential for life have formed around stars in other parts of the Universe.

In this chapter we take an in-depth look at the structure and composition of Earth. We start by looking at how the abundant elements within our planet have combined into rocks, what determines the properties of these rocks, and how these properties are key to many important properties of Earth. We will look at the different rock types and learn how geologists can tell them apart. Radioactive substances play an important role in geology and we will see how they provide a means of determining the age of Earth and its rocks. Moving on to the large-scale structure, we learn how Earth differentiated into an iron core, mantle, and crust, and how geologists use earthquakes to determine the mineralogical and chemical composition of the inner Earth.

8.1 Composition of Earth

Geologists have studied the composition of Earth's surface rocks and probed Earth's interior. Their findings reveal that the major ingredients of our planet are among the most common elements found throughout the Universe. Table 8.1 lists the most abundant elements in the Universe, deleting those elements (e.g., hydrogen, helium, carbon, and the rare gases like argon and neon) that are found primarily in gaseous form in space. These elements would not have condensed into solids at the high temperatures of the inner solar nebula, and would have been driven out of the inner Solar System by the subsequent flare-up of the young Sun. Therefore, we would not expect to find them in abundance in earthlike planets.

The eight most abundant elements that make up Earth are also listed in the table. Comparison of the eight most abundant, non-volatile elements in the Universe with the list of the most abundant elements in Earth reveals a striking agreement. The lists differ somewhat in relative abundances (iron is the most abundant element in Earth, for example, but is only the fifth most abundant element in the Universe), but otherwise the agreement is excellent once the volatiles have been removed.

The fact that Earth is largely made up of some of the most common elements in the Universe supports the idea that whenever a family of planets forms around a star anywhere

Previous page: The planet Earth, as seen from the Apollo 17 spacecraft.
NASA

Table 8.1 *Most abundant elements (listed in order of decreasing abundance)*

Most abundant elements in the Universe (minus volatiles)	Most abundant elements in Earth
Oxygen (O)	Iron
Silicon (Si)	Oxygen
Magnesium (Mg)	Silicon
Sulfur (S)	Magnesium
Iron (Fe)	Nickel
Aluminum (Al)	Sulfur
Calcium (Ca)	Aluminum
Nickel (Ni)	Calcium

in the Universe, it will probably contain some planets made of the same materials as Earth, i.e., earthlike planets. This is an important conclusion because there is evidence that trillions of solar systems may exist in the observable Universe.

8.2 The chemical bond

The earthlike planets are rocky bodies. How do the abundant elements in Earth combine to make the substance we call rock? This is a question of chemistry: What chemical compounds are likely to be formed out of the basic ingredients of an earthlike planet? The answer involves the laws of the chemical bond that govern the way in which separate atoms combine to form solid substances.

8.2.1 Laws of the chemical bond

The way in which chemical elements combine depends upon the arrangement of electrons in orbit around the atomic nucleus in each element. In the atomic world, electrons behave in accordance with a special set of rules:

Rule 1
The electrons in an atom can only circle the nucleus in certain specific orbits called "allowed" orbits. In the hydrogen atom, for example, the first allowed orbit for the electron has a radius of

0.53 Å. (The angstrom, written Å, is equal to 10^{-8} centimeters. It is a convenient unit for expressing sizes of atoms.) No hydrogen atom exists with an electron in an orbit with a radius of 0.4 Å or 0.6 Å; the orbit size has to be 0.53 Å.

Certain other allowed orbits exist for the electron in the hydrogen atom; for example, beyond 0.53 Å there is another allowed orbit with a radius of 2.12 Å. But no hydrogen atom is ever found in which the electron is in an orbit with a radius of 1.8 Å or 2.5 Å – only 2.12 Å.

The atoms of other elements also have allowed orbits for their electrons. These are different, in general, from the allowed orbits for the hydrogen atom. Each element has its own special set of allowed orbits that are unique to that particular atom.

Rule 2

Each allowed orbit in an atom can hold only a limited number of electrons. The innermost allowed orbit can hold up to two electrons (but no more than two). The second allowed orbit can hold up to eight electrons (but no more than eight). The third orbit can also hold up to eight electrons. The fourth orbit can hold up to eighteen electrons; and so on. If an orbit contains the maximum number of electrons it is allowed to have by this rule, the orbit is said to be full or closed (Figure 8.1).

Rule 3

Atoms tend to form closed electron orbits. If, for example, the number of electrons in an atom is one less than the number needed to fill or close the atom's outermost orbit, that atom

FIGURE 8.1 The electron orbital configurations of the first 18 elements in the periodic table of the elements.

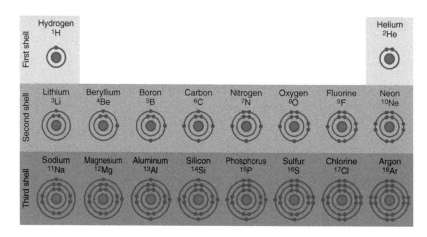

will tend to pick up an electron from any other atom that may be nearby, and close its outer orbit in that way. But, of course, the other atom will not let go of one of its electrons, because they are held to it by the electrical force of attraction to the nucleus. Thus, the two atoms tug at the same electron. In tugging at the same electron, they are tied to one another. They have formed a chemical bond.

As another example, suppose an atom has one electron more than the number needed to close its outermost orbit. Since atoms like to have closed orbits by Rule 3, that atom will not hold on to this last electron very tightly. It will share it readily with any other atom passing by. But as before, it cannot give up the outermost electron entirely, because that electron is still tied to its nucleus. So, once again, the two atoms may end up tugging at the same electron. In so doing, they are tied to one another. This is another way in which a chemical bond can be formed.

If an atom has precisely the number of electrons needed to close or fill all its orbits, it tends neither to pick up electrons from other atoms, nor to give up electrons to those atoms. In other words, it is relatively indifferent to the presence of neighboring atoms, and is not likely to form chemical compounds. It is thus chemically inert or inactive.

8.2.2 Examples of chemically inactive and active elements

Neon (Ne) has a total of ten electrons. Two electrons are in the first or innermost orbit of the neon atom, closing that orbit. The next eight electrons go into the second orbit, closing that one also. Since both electron orbits in an atom of neon are filled or closed, this substance is not prone to take an electron from another atom, nor to give one up to another atom. It is chemically inactive. It is called a *noble gas* because it does not like to form a union with any other atom.

Oxygen (O), on the other hand, is chemically active. Why is that? The answer is that oxygen has eight electrons. Two of these go into its first orbit, and close it. That leaves six to go into the second orbit. But the second orbit can hold eight electrons. With only six electrons in it, the orbit is not closed; it needs two more electrons to close it. Therefore, by

Rule 3 of the atom, oxygen tries to grab two electrons from any other atom nearby, in order to close that second orbit.

That nearby atom could be silicon (Si), for example. The silicon atom has fourteen electrons. Two are in the first orbit. Eight are in the second orbit, and fill it. That leaves four more electrons, which are outside the second orbit. If an oxygen atom comes close to a silicon atom, it will attract those outermost two electrons and tend to hold on to them, because they can partly fill its second orbit. But the oxygen atom cannot tear the two electrons completely away from the silicon atom; those electrons are held to the silicon atom by the positive charge on the silicon nucleus. Although bound somewhat loosely to the silicon atom, they are still bound.

But one oxygen atom will only account for two of the four electrons available in the outermost orbit of the silicon atom. Two oxygen atoms, however, would take up all four electrons in the outer orbit of silicon. Thus, a likely chemical combination is two atoms of oxygen joined to one atom of silicon.

These three atoms form the compound silicon dioxide (SiO_2). In this compound, each oxygen atom tries to take two electrons away from the silicon atom. The silicon atom cannot give up these electrons, because they are held to its nucleus. The result is that both atoms tug at the same electrons, and in so doing, they are tied to one another in a chemical bond.

The electron structures of the various atoms are the key to the understanding of the chemical bonds. Consider the elements oxygen and hydrogen. A natural compound formed from these two elements would be a molecule composed of one atom of oxygen joined to two atoms of hydrogen. Why? Because the oxygen can share an electron with each of the two hydrogen atoms, and this way gain partial possession of the two electrons it needs to fill its outermost orbit. The molecule that results from these bonds is, of course, water (H_2O).

Carbon (C) is another example. The carbon atom has six electrons – two electrons filling the first orbit, and four electrons outside that in the second orbit. Since the second orbit can hold eight electrons, it is only half filled. Four more electrons would fill it.

Thus, carbon can form a chemical combination with the hydrogen atom, which can share an electron with it and make a contribution toward filling that half-filled second orbit in carbon. Four hydrogen (H) atoms, arranged evenly around the carbon atom, would contribute the four electrons needed to fill the second orbit in the carbon atom entirely, and would make a particularly strong bond. This union of a carbon atom with four hydrogen atoms is, in fact, the molecule methane (CH_4).

But note that the four electrons in the second orbit of carbon can also be regarded as four excess electrons outside a closed first orbit. That means carbon can form compounds with atoms that have empty places in their outer orbits, and need electrons to fill these places.

For example, carbon could form a compound with oxygen, in which the carbon atom contributes two electrons to fill the outer orbit of the oxygen atom. A particularly strong bond would be one in which one carbon atom joined with two oxygen atoms, using up all four excess electrons, and completely filling the orbits of both oxygen atoms. The compound that results from this union is the familiar molecule carbon dioxide (CO_2). Chemically speaking, the carbon atom has a dual personality.

8.2.3 The common Earth elements and their electron configurations

The eight most common elements in Earth's crust are shown in Table 8.2. These elements make up the bulk of the rocks near Earth's surface. Sodium and potassium are more abundant in the crust than in Earth as a whole because they enter into the structure of a very common mineral in Earth's crust called feldspar.

The table also gives the number of electrons in the outermost orbit of each element and the size of its atom. (The table actually lists the size of the atomic ion formed by adding electrons to, or subtracting them from, an atom so as to make a closed orbit configuration. Since atoms tend to form compounds in such a way that each approaches a closed-orbit configuration, the size of the ion is more relevant here than

Table 8.2 *Number of electrons, orbit structure, and size of the eight most common elements in Earth's crust*

Element	Total number of electrons	Orbit structure	Ionic size (radius in Å)
Oxygen (O)	8	needs 2 electrons to close orbit	1.3 (large)
Silicon (Si)	14	4 excess electrons	0.4 (small)
Magnesium (Mg)	12	2 excess electrons	0.7 (medium)
Iron (Fe)	26	(complicated)	0.7 (medium)
Aluminum (Al)	12	3 excess electrons	0.5 (small)
Calcium (Ca)	20	2 excess electrons	1.0 (large)
Potassium (K)	19	1 excess electron	1.3 (large)
Sodium (Na)	11	1 excess electron	1.0 (large)

$1 \text{ Å} = 10^{-8} \text{ cm}$

the size of the isolated atom.) The electron configuration and the size of an atomic ion together determine the particular compounds it is likely to form.

8.2.4 Compounds of silicon and oxygen: The silicate minerals

Compounds of oxygen and silicon, called *silicate minerals*, are of particular importance for the study of Earth. Consider the chemical properties of silicon. Of the fourteen electrons in the silicon atom, two electrons fill the first orbit, eight fill the second orbit, and four go into the third orbit and partly fill it. These four electrons in the third orbit can be shared with other nearby atoms.

A particularly suitable chemical partner for silicon would be an atom that needs some electrons to fill its outermost orbit. Oxygen is such an atom as we have seen. Since oxygen needs two electrons to close its orbit, and silicon has four that are available, a likely chemical compound of silicon and oxygen would be two atoms of oxygen joined to one atom of silicon. This compound is abundant in Earth; it is silicon dioxide (SiO_2), also called silica or the mineral quartz.

Quartz is just one of a number of silicate minerals that make up the body of Earth. In addition to silicon and oxygen, the silicate minerals contain other common elements such as iron (Fe) and magnesium (Mg). The silicate minerals are sometimes also called the rock-forming minerals because they are the substances from which most kinds of common rocks are made.

The properties of the silicate minerals are the key to many important properties of Earth. The formation of these minerals depends on a second factor that influences the structure and composition of a chemical compound, in addition to the electron orbit structure of its atoms. This second factor is the size of each atom (or, more precisely, its atomic ion).

Consider silicon and oxygen, which make up the compound silicon dioxide. Silicon has a radius of 0.4 Å. The radius of oxygen, on the other hand, is 1.3 Å. Drawing the atoms to scale, the compound silicon dioxide should resemble Figure 8.2.

However, as Figure 8.2 indicates, because the oxygen atoms are so large compared to the silicon atom, they are in contact with the silicon atom only over a relatively small fraction of their surface areas. This tends to weaken the force of the chemical bond created by the sharing of outer electrons between the two kinds of atoms.

But a different arrangement of silicon and oxygen atoms is possible, in which the degree of contact between neighboring atoms – and, therefore, the strength of the chemical attraction – is much greater. Suppose we place four oxygen atoms around the relatively small silicon atom, so that it sits inside a nest of oxygens. In this arrangement, the compound resembles a pyramid, with one oxygen atom at each corner of the pyramid, and the silicon atom in the center of the pyramid. If lines are drawn connecting the centers of the oxygen atoms, the resultant figure is a tetrahedron – a four-sided figure, each side of which is a triangle. In silicate minerals, this arrangement is called the *silica tetrahedron* (Figure 8.3).

The arrangement of four oxygens around one silicon provides the maximum degree of contact possible between the silicon atom and the oxygen atoms. If we tried to place more than four oxygen atoms around one silicon atom, they would not be as close to the silicon atom, or as tightly bound to it, as they are in the tetrahedral structure. Because the structure of the tetrahedron creates a strong bond between oxygen and silicon atoms, the silica tetrahedron is the basic building block out of which all silicate minerals are made.

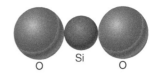

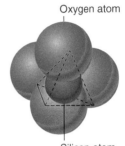

FIGURE 8.2 A possible arrangement of atoms for silicon dioxide (SiO_2).

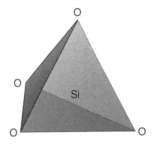

FIGURE 8.3 Four oxygen atoms surrounding a silicon atom make up the basic unit of silicate mineral structure. The dashed lines which connect the centers of the atoms form a tetrahedron – that is, a four-sided figure in which each side is a triangle. The figure on the right shows the tetrahedron more clearly.

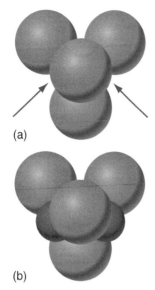

(a)

(b)

FIGURE 8.4 (a) The silica
tetrahedron. The silicon
atom is concealed by one of
the oxygen atoms in this
view, and is not visible. The
arrows point to two of the
hollow spaces or "holes" in
the faces of the tetrahedron.
(b) If iron (Fe) or magnesium
(Mg) atoms are placed in
these "holes," the result is
the basic structure of the
mineral olivine. Since a
tetrahedron has four faces,
there are actually four
"holes" in each
tetrahedron. In an actual
olivine crystal, all four
holes are filled with iron or
magnesium atoms.
However, each iron or
magnesium atom is shared
between two adjoining
tetrahedra, so that there is
an average of two iron or
magnesium atoms per
tetrahedron.

But the structure of the silica tetrahedron creates a problem, chemically speaking. Oxygen lacks two electrons to close its second orbit, so the four oxygen atoms in the tetrahedron require a total of eight electrons to close their orbits. The silicon atom in the center only has four electrons to share with them, which means the oxygen atoms still have a total of four unfilled places in their outer orbits. Therefore, this combination – the silica tetrahedron, or SiO_4 – will tend to form a chemical union with any other atoms around having electrons outside closed orbits and available for sharing.

Which atoms are likely candidates for sharing electrons with the silica tetrahedron? Iron, magnesium, aluminum, calcium, potassium, and sodium are suitable because all have electrons that are readily shared with oxygen.

Iron, for example, has two electrons to share. Two iron atoms, added to a tetrahedron, would contribute four electrons in all, filling the orbits of the oxygen atoms in that tetrahedron. Furthermore, since the iron atoms are quite small, they fit nicely into the hollow spaces in the triangular faces of the tetrahedron. Thus, the degree of contact is very great, which makes this combination – a silica tetrahedron plus two iron atoms – a very tightly bound unit. A silica tetrahedron with two iron atoms added is, in fact, the basic building block of an important silicate mineral called *olivine*. Olivine is one of the most tightly bound of all silicate minerals (Figure 8.4).

Magnesium has the same number of electrons to share and ionic size as iron, and should fit into the structure of the silica tetrahedron equally well, forming the same kind of silicate mineral. In fact, it should be possible to add to the silica tetrahedron either two iron atoms, or two magnesium atoms, or one iron atom plus one magnesium atom. All three compounds are similar in their properties; all three make up varieties of the same silicate mineral, *olivine*.

One feature of this silicate mineral remains to be explained. A solid chunk of olivine – a piece you can hold in your hand – consists of many silica tetrahedra bound together. What force binds these tetrahedra?

The bonding between tetrahedra comes about through the fact that one iron or magnesium atom nests in each of the four triangular faces of each tetrahedron, but every iron or magnesium atom is shared between two adjoining tetrahedra. Each iron or magnesium atom, sharing its electrons with two adjoining tetrahedra, ties those tetrahedra together. Many tetrahedra tied together in this way make a crystal of olivine.

Because in olivine all four "holes" in the four faces of the silica tetrahedron are filled with iron or magnesium atoms, this mineral is very dense. The high density is due partly to the fact that iron and magnesium are relatively heavy elements (atomic weights (protons + neutrons in the nucleus) of 56 and 26, respectively), and partly to the fact that every open space in the structure is filled with an atom. Olivine is, in fact, the densest of the common silicate minerals, a fact that will become important when we look at the structure of Earth.

Because iron and magnesium atoms are small, they fit into the "holes" in the faces of the silica tetrahedra easily, without forcing the tetrahedra apart, which would weaken the bonds. This means the chemical bonds in olivine are strong – stronger, in fact, than in any other common silicate mineral. The great strength of the bonds in olivine means, in turn, that very high temperatures are required to melt this mineral. (Melting a mineral means raising its temperature until the vibrations of the atoms break the bonds that hold the mineral together. The stronger the bonds, the higher the temperature required to break them.) The melting-point temperature of olivine is, in fact, higher than that of any other common mineral.

A silicate mineral can also contain calcium, which, like iron and magnesium, has two electrons to share and therefore can combine with silica tetrahedra. However, the calcium atom, being more than twice the size of an iron or magnesium atom, forces the silica tetrahedra apart, so that the resultant silicate mineral is not as tightly bound as olivine. As a consequence, silicate minerals with calcium built into their structure do not have as high a melting-point temperature as olivine.

In another common type of silicate mineral, adjoining tetrahedra share an oxygen atom as a common vertex or corner. Figure 8.5(a) shows how two tetrahedra can have one oxygen atom as a common vertex. Many tetrahedra can be linked together in this way, sharing oxygen atoms as common corners, to form a long chain (Figure 8.5(b)). Each

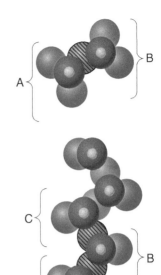

FIGURE 8.5 (a) The structure of the mineral pyroxene. Two adjoining silica tetrahedra are shown, marked A and B. They share a common oxygen "corner." The diagonal-line shading indicates the shared oxygen atom. Smaller iron or magnesium atoms sit on top of the tetrahedra. (b) The structure of pyroxene. The figure shows how a long chain of silica tetrahedra can be formed from many silica tetrahedra. Each shares an oxygen "corner" with the tetrahedra on either side of it. The diagonal-line shading indicates two of the shared oxygen atoms in the chain. These atoms belong to tetrahedron B, but they also belong to tetrahedra A and C, above and below B in the chain. Iron and magnesium atoms join the chain of tetrahedra to other similar chains.

silica tetrahedron in the chain shares a vertex with two tetrahedra – one on each side. That means that each tetrahedron has two oxygen "corners" that are shared with adjoining tetrahedra on either side of it in the chain, and two "corners" that are not shared.

In this common silicate mineral, the three oxygen atoms that go with each silicon atom lack a total of six electrons to fill their orbits. The silicon atom has four electrons to share with them. Hence this combination, SiO_3, needs *two* more electrons to fill the oxygen orbits. One iron or magnesium atom, added to the SiO_3 structure, would contribute two electrons and fill those orbits. The resultant silicate mineral, built out of long chains of tetrahedra, each sharing a corner with its neighbors, and with one iron or magnesium atom per tetrahedron, is called *pyroxene*.

Pyroxene, having only one iron or magnesium atom per tetrahedron, is less dense than olivine. Also, having fewer irons and magnesiums than olivine, it has fewer chemical bonds tying neighboring tetrahedra together. This means that its atoms are not as tightly bound as in olivine, and, therefore, pyroxene melts more easily – its melting-point temperature is lower.

In a third type of silicate mineral, each silica tetrahedron shares all four oxygen "corners" with adjoining tetrahedra, instead of sharing just two, as in the case of pyroxene. In this mineral, every oxygen atom is shared between two silicon atoms. Thus, the chemical formula of the mineral is SiO_2. This is the mineral *quartz*.

Quartz, with no iron or magnesium atoms, and, in fact, no atoms of any kind to fit into the spaces between adjoining tetrahedra, has a very open structure with many "holes." For that reason it has the lowest density among the common silicate minerals. Also, because quartz has no atoms to fill in the holes in the structure, it lacks the additional bonds that exist in olivine and, to a lesser degree, in pyroxene. As a consequence, quartz can be melted – that is, broken into separate tetrahedra – much more easily than olivine and pyroxene.

Still another variation on the tetrahedral structure leads to the most common mineral in Earth's crust. This mineral has aluminum built into its structure. According to Table 8.2, the aluminum ion is nearly as small as the silicon ion, and therefore can fit almost as well as silicon into the hollow space at the center of a tetrahedron formed by four oxygen atoms. If a molten mass of rock materials has a large amount of aluminum, some aluminum atoms will wind up in the centers of tetrahedra as the mass cools, replacing silicon atoms.

But the aluminum atom has three electrons outside a closed orbit. Therefore, a tetrahedron with an aluminum atom at the center has one less electron to share with the surrounding oxygen atoms than in the case of a tetrahedron with a silicon atom at the center. To fill all the orbits in these oxygen atoms will require one additional electron. If an atom of potassium, sodium, or calcium is present in the molten mass, it can supply the extra electron.

So now we have a new type of mineral in which some tetrahedra have aluminum atoms in their centers instead of silicon atoms. A mineral of this type, in which aluminum substitutes for silicon in the centers of some tetrahedra, and the electrons missing from those tetrahedra are contributed by atoms of potassium, sodium, or calcium, is called *feldspar*.

In fact, feldspar is the most common mineral in Earth's crust. It comes in two varieties: *orthoclase feldspar*, in which potassium atoms contribute the extra electrons, and *plagioclase feldspar*, in which the extra electrons are contributed by sodium and calcium. Because the potassium atom is larger than atoms of sodium or calcium, it distends the structure of orthoclase feldspar, so that this type of feldspar tends to have a somewhat lower density and lower melting-point temperature than plagioclase feldspar.

8.3 Rocks

Most rocks are mixtures of silicate minerals. If the rock has been formed by the cooling and solidification of a molten mixture of silicate minerals, it is called an *igneous rock*. The other major types of rocks are: *sedimentary rocks*, composed of fragments of pre-existing rocks that have been worn away by the forces of running water and wind and then cemented or compacted together again; and *metamorphic rocks*, formed when sedimentary or igneous rocks are subjected to high temperature and pressure, but not melted.

For igneous rocks, the elements present in the original molten mass called *magma* determine which silicate minerals crystallize out of the mass as it cools, and thus the type of rock that forms. For example, magma rich in iron and magnesium will crystallize into silicate minerals containing an abundance of these elements (i.e., olivine and pyroxene). On the other hand, magma that is initially poor in iron and magnesium will crystallize instead mainly into minerals that contain little or no iron or magnesium (i.e., quartz and feldspar).

Igneous rocks are also characterized by the size of the individual mineral crystals in the rock. If a rock forms from magma that has cooled very slowly, the crystals will be large. If the magma cools very rapidly, the crystals in the resultant rock will be small.

The reason is that as the magma cools, microscopically small grains of silicate minerals begin to form in the melt. First two or three silica tetrahedra stick together; then additional tetrahedra stick to them; and so on. With the passage of time, more and more atoms stick to this tiny nucleus of a crystal, and it grows larger. If the cooling is rapid, growth of the crystals is interrupted after a short time, and the entire molten mass solidifies into a rock before any crystals have time to grow to a large size. All the crystals in the rock then are small, and the rock has a fine-grained texture. But if the melt cools slowly, the grains have time to grow into large crystals, and the rock has a coarse-grained texture.

The size of the grains in an igneous rock contains important information about the origin of the rock. For example, magma cools quickly if it erupts from a volcano because it loses its heat rapidly to the atmosphere. In fact, it can congeal into solid rock in as short a time as a few minutes. Consequently, if a geologist picks up a piece of fine-grained igneous rock, the geologist knows it has probably been formed in a volcanic eruption.

But some magmas never reach Earth's surface. They may sit in a subterranean magma chamber cooling very slowly while insulated by the cooler rocks around them. These subterranean pockets of magma can take thousands of years to solidify completely, providing enough time for their crystals to grow to a large size. If an igneous rock has a coarse-grained texture, the geologist knows that it formed deep within Earth.

8.4 Melting of Earth

Earth, like the Moon, must have experienced two periods of melting; a first melting as a result of bombardment during its formation and a second melting as a result of radioactive heating produced by the decay of uranium (U), thorium (Th) and an unstable form of the element potassium (K). These radioactive elements were included in the cloud of materials out of which the Solar System formed, and were incorporated into Earth and the other earthlike planets when they accreted from that cloud of debris. The large nuclei of the radioactive elements disintegrate spontaneously into smaller nuclei. In doing so, they throw off fragments of nuclear matter. The fragments collide with the atoms around them and transfer energy to these atoms, heating the surrounding rock (Figure 8.6). Eventually the temperature can reach the melting point of rock as a result of this radioactive heating.

8.4.1 Properties of radioactive substances
Radioactive substances play a very important role in geology, not only because they heat the interior of Earth, but they also provide a means of determining the age of Earth and its rocks.

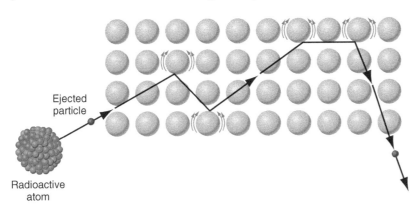

Ejected particle

Radioactive atom

FIGURE 8.6 The heating effect of radioactive elements. A particle ejected from a radioactive atom strikes atoms of the surrounding rock and transfers energy to them.

The atoms of the radioactive substances, such as uranium, are locked into the crystal structure of the minerals that make up the body of Earth. After a time they decay, i.e., break up into smaller nuclei. Any particular atom of uranium, for example, may decay in the next minute, or in 10 million years from now, or a billion years from now; one never knows exactly when a particular uranium nucleus will disintegrate. However, physicists have found that if they study a sample of uranium containing a large number of uranium nuclei, after 4.5 billion years half those nuclei will have decayed. Four-and-one-half billion years is called the *half-life* of uranium. (The close agreement between this number and the 4.6-billion-year age of the Solar System is a coincidence). This means that if 1000 uranium atoms exist at a given moment, after 4.5 billion years only 500 of those atoms will be left. The other 500 atoms will have decayed.

The amount of heat generated as a by-product of the decay of just one uranium atom is very small, enough to heat one cubic centimeter of rock by about one million trillionth of a degree in a year. But one cubic centimeter of a typical rock may contain many trillions of atoms of uranium and other radioactive elements. Calculations indicate that the small amount of radioactive heat, accumulating from the decay of these trillions of atoms, would have been enough to raise large parts of Earth's interior to the melting point of rock within a few hundred million years after the formation of the planet.

8.4.2 How radioactivity is used to determine the "age" of a rock

Now we are ready to apply the properties of radioactive substances to a measurement of the ages of rocks. But first we must still settle one question. We have discussed rocks of various ages on Earth and the Moon, but what exactly do we mean by the age of a rock? After all, in one sense, the atoms in a rock are older than the Solar System. Most were manufactured in stars that lived and died before the Solar System existed. The individual neutrons and protons in the nuclei of those atoms are even older than the atoms; they are the primordial stuff of the Universe, formed 14 billion years ago.

What, then, is meant by the "age" of a rock made from those ancient atoms? The answer is that at some point in the history of Earth, the atoms in the rock congealed to form the piece of solid matter (a rock) that we hold in our hand today. The existence of the rock dates back to the moment when that happened.

8.4.3 Dating a rock by its lead/uranium ratio

How do we measure the age of the rock, defined in this way? One way is to examine the history of the radioactive substances in the rock. Nearly every rock has some of these substances – uranium and other radioactive elements – in it. Usually the amount is small – perhaps one atom of uranium for every million atoms of the abundant elements like silicon or iron. In the course of time, the isotope of uranium (U) in the rock decays atom by atom to an isotope of the element lead (Pb). This happens at a known rate (as we have seen, half decays in 4.5 billion years). Therefore, if the rock is very old, it will have a relatively large amount of lead in it, and very little uranium, because most of the uranium will have changed to lead. If the rock is young, it will have a relatively large amount of uranium, but not yet very much lead.

Knowing the rate at which uranium decays to lead, and measuring the ratio of lead to uranium in a particular rock, we can calculate how old that rock is. Thus, radioactivity gives us a clock for determining the rock's age.

It is important to note that the so-called "radioactive clock" only starts ticking after a rock has congealed into a solid chunk of material. Only then are the atoms of lead and uranium frozen into the rock, so that no lead atoms or uranium atoms can get in or out. Only if that is true will the ratio of lead to uranium atoms in the rock tells us its age. Two complications can affect the accuracy of the lead/uranium "radioactive clock." First, lead produced earlier by the decay of uranium might be present in the magma before it congealed into a rock. This excess lead would cause the measured "age" of the rock to seem older than it actually is. By carefully choosing a mineral in which lead is not a normal constituent, however, we can ensure that all or nearly all lead atoms found in that mineral were produced by the uranium atoms contained within the material at the time it solidified, and that no lead atoms were inherited from its past.

The second problem is that some lead atoms are "non-radiogenic," that is, they were not formed by the decay of uranium atoms at all, but were produced, along with other heavy elements, in supernova explosions before the Solar System existed. These non-radiogenic lead atoms may be distinguished from radiogenic lead atoms by the fact that non-radiogenic lead atoms commonly have 204 nuclear particles (82 protons, 122 neutrons) whereas lead atoms produced by the decay of uranium, as described above, have 206 nuclear particles (82 protons, 124 neutrons), or 207 or 208.

Other radioactive elements can be used for the measurement of the age of a rock. For example, one variety of the element potassium is radioactive and decays at a known rate to form the element argon. Therefore, a rock's age can be determined by measuring the ratio of argon to potassium. This measurement can be used independently, and as a check on the results obtained from the lead/uranium ratio.

Dating of rocks by these methods has revealed that the oldest rocks on Earth are 3.8 to 3.9 billion years old. However, the oldest objects in the Solar System that have been dated – including many meteorites and some rocks from the lunar highlands – are up to 4.6 billion years old.

8.5 Differentiation of Earth

Earth formed rapidly out of bits of rock and metal that collided and stuck together in a process called accretion. The heat liberated in the collisions caused the temperature of the growing planet to rise. As the temperature rose, iron metal was the first substance to melt. Once molten, the iron, being relatively dense, began to sink toward the center of the planet to form Earth's *core*. The iron took with it other relatively dense elements that have an affinity for iron, such as nickel, gold, and platinum.

As the iron moved toward Earth's center, its flow was impeded by frictional resistance, which heated the surrounding rock. In this way, Earth's internal temperatures rose dramatically and silicate minerals began to melt as well. As the dense iron metal accumulated at the center of Earth, the less dense molten silicates rose upward to form a jacket of rock around the iron core. This jacket of rock, lying outside the iron core and surrounding it, is called the *mantle*. The mantle may have been largely molten during the accretion of Earth, but subsequently cooled as the bombardment tapered off. With the passage of time, however, as a result of radioactive decay, the temperature in the mantle reached the melting point of rock and the materials in Earth's interior began to melt a second time.

Now a very important property of silicate minerals enters the picture. Consider molten rock that is cooling and solidifying. As the mass of molten rock cools, each mineral will freeze out and solidify at a different temperature. Furthermore, the different minerals also have different densities. In general, minerals that freeze at the highest temperatures have the highest densities. The mineral olivine has the highest freezing-point temperature. Therefore, olivine would have been the first mineral to crystallize out of the melt. Olivine is also denser than the other common silicate minerals, and therefore, as soon as it crystallized, the solid crystals would tend to sink to the bottom of the melt. Accordingly, the lowest part of the mantle must be rich in olivine.

The mineral pyroxene has the next highest freezing-point temperature. Therefore, in later stages of cooling, pyroxene will solidify. Pyroxene is also the densest mineral after olivine. Therefore, above the region with a high concentration of olivine there will be a region with a relatively high concentration of pyroxene.

The last minerals to freeze out, and also the lowest density minerals, are feldspar and quartz. Accordingly, these minerals will be found in abundance at the shallowest depths, in Earth's *crust*.

According to this picture of Earth's large-scale structure, the interior of Earth is made up of a core, a mantle, and a crust. The core of iron, largely molten but with a solid inner core, occupies the central region of Earth. (Recent studies show that Earth's inner core is rotating faster than the crust and mantle.)

FIGURE 8.7 Structure of the differentiated Earth, showing the major zones in Earth's interior – the core, mantle, and crust.

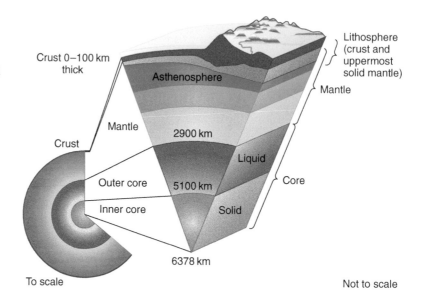

Surrounding it is a mantle of silicate minerals. Within the mantle there is a gradual change in composition. The most iron-rich and magnesium-rich minerals are concentrated in the lower mantle, and minerals less rich in iron and magnesium are in the upper mantle. Near the surface, a sharp transition occurs to an outer layer – Earth's crust.

The process by which Earth is partitioned into a core, a mantle, and a crust, and the various silicate minerals concentrated at different depths within the mantle, is called *differentiation*. Figure 8.7 shows the differentiated structure of Earth's interior. The molten iron outer core is approximately 2000 miles (3200 kilometers) in radius. Surrounding the core is a mantle of rock approximately 2000 miles (3200 kilometers) thick, and extending nearly to the surface of Earth. On top of the mantle is the crust, averaging 12 miles (19 kilometers) in thickness, but whose thickness varies from as little as 3 miles (4.8 kilometers) under most parts of the ocean to 35 miles (56 kilometers) under some parts of the continents.

An inner core of solid iron, with a radius of 1000 miles (1600 kilometers), sits within Earth's outer core of molten iron (Figure 8.7). This solid inner core is explained by the fact that, at the enormous pressures prevailing near the center of Earth, iron atoms are locked into the regular positions of a lattice structure – i.e., a solid – in spite of the very high temperatures prevailing there.

The composition of Earth varies with depth in another important respect. Certain elements, whose ions have a very large size, do not fit into the compact structure of the silicate minerals found at great depths. There is no room for them in the structure of these very dense minerals. Consequently these large-diameter atoms are "squeezed" up toward the surface, and become concentrated in the minerals of the upper mantle and the crust, which have a more open crystal structure.

Calcium, sodium, and potassium all have large atomic diameters and are concentrated in the crust by this process. The atoms of the element uranium also are very large. Consequently, uranium is also concentrated in the rocks of Earth's crust. Typical crustal rocks have between 50 and 200 times more uranium than the rocks of the mantle.

8.6 Earthquakes and Earth's interior

How do geologists know the details of the internal structure of Earth? To probe Earth's interior, geologists utilize the energy released by earthquakes. *Seismic waves* from an earthquake propagate through the bulk of Earth and at distant locations they can be recorded by instruments called *seismometers*, which amplify and record subtle shifts of the ground caused by arriving waves.

A network of seismic observation stations has been established around the globe to record earthquake events. By comparing the signals received by different stations, one can determine the precise location of an earthquake, and the time that it occurred. The key to determining the interior structure of Earth involves precise measurements of the travel time for the waves arriving at distant observatories. Earthquake waves travel faster when they pass through denser materials, and they arrive at distant stations bearing information about the structures they encountered along the way.

Earthquakes generate two very different kinds of waves, called *P-waves* and *S-waves*. P-waves are simple *pressure waves*; as the wave passes, material is alternately compressed and decompressed, as shown in Figure 8.8(a). Sound waves in air or water work the same way. S-waves are peculiar to rigid solids, and the

FIGURE 8.8 The nature of P-waves and S-waves in a rigid solid. (a) Compressional (p) waves. (b) Shear (S) waves.

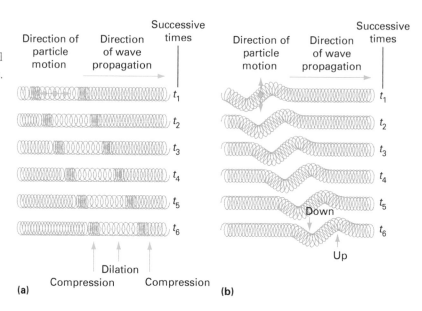

way they propagate is shown in Figure 8.8(b). The passing wave causes material to shift back and forth *laterally*, transverse to the path of the wave. Material is not compressed by such waves; instead, it moves sideways laterally as the wave passes.

The distinctive property of S-waves is that *they cannot propagate in a liquid medium*. If you apply a lateral force to a liquid, it simply flows and the stress is relieved instantly. P-waves and S-waves also propagate at different speeds, with P-waves moving quite a bit faster than S-waves. (P-wave velocities in Earth range from 6 to 13 kilometers (4 to 8 miles) per second; S-wave velocities range from 3.5 to 7.5 kilometers (2.2 to 4.7 miles) per second.)

That means the S-waves will take longer to reach a distant seismic station and can be clearly distinguished from arriving P-waves. The paths followed by waves from an earthquake as they pass through Earth's mantle are refracted or bent because the density of the material increases with depth, causing the waves to increase in velocity. P-waves and S-waves follow roughly the same paths, as long as the material they traverse is solid, but they travel at different velocities (Figure 8.9).

One of the first discoveries made by seismologists was that S-waves from an earthquake could not be detected by stations lying in a wide zone on the opposite side of Earth, called the S-wave *shadow zone*. Because the transmission of the S-waves was blocked, they concluded that a liquid core with a radius of

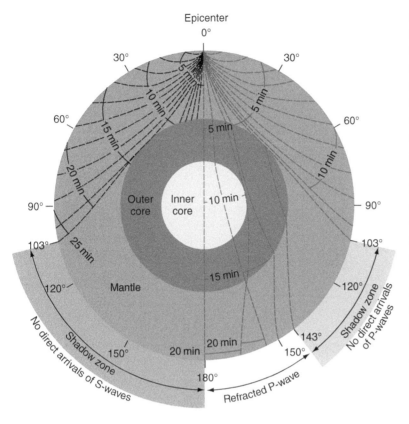

Epicenter
0°

30°   30°

60°   60°

5 min

Outer   Inner 10 min
core   core

90°   90°

103°   103°

25 min

Mantle   120°

120°

15 min

143°

No direct arrivals of S-waves   150°

20 min   20 min
150°

180°

Refracted P-wave

Shadow zone
No direct arrivals
of P-waves

Shadow zone

FIGURE 8.9 Paths of P-waves and S-waves from an earthquake. S-waves that encounter the molten outer core dissipate and cannot reach observation stations on the other side of Earth further than about 103 degrees from the earthquake epicenter (the shadow zone for S-waves). P-waves can pass through both outer and inner cores, but their velocities change abruptly at each boundary, causing refraction and reflection of the waves. A shadow zone for P-waves exists between 103 and 143 degrees from the earthquake epicenter. Travel times for P-waves and S-waves within Earth are also shown.

about 2200 miles (3500 kilometers) must exist (Figure 8.9). Later studies of P-waves, which do pass through the core, showed that there is a smaller solid core inside the liquid core. S-waves are observed to propagate through all parts of the mantle, so we know that Earth's mantle is solid, or at least semi-solid, down to the boundary with the liquid iron core.

By performing laboratory experiments on the speeds of propagation of sound waves through various minerals, and comparing these values with those recorded for seismic waves passing through Earth, seismologists were able to determine the mineralogical and chemical composition of the inner Earth. The various zones are clearly delineated (Figure 8.10). Some correspond to changes in chemical composition, such as at the core–mantle boundary. Within the mantle, however, as pressure and temperature increase with depth, the dominant mineral olivine can undergo a transformation from its normal crystalline form to another form that is denser. This unusually dense mineral with the same composition as

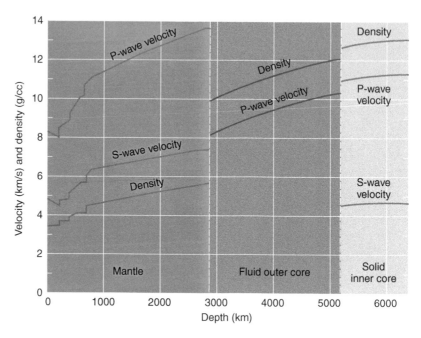

FIGURE 8.10 Density at different depths in Earth, as inferred from seismic studies of velocities at which P-waves and S-waves propagate. The wave velocities in the mantle provide evidence for a low-velocity zone at around 60 miles (100 kilometers) in depth, and sudden increases in density at about 250 miles (400 kilometers) and 420 miles (670 kilometers) deep.

olivine is called *perovskite*. Because the two minerals have different densities and different wave propagation character-istics, the boundary between them shows up as a discontinu-ity in the velocity of the seismic waves at a depth of about 240 miles (400 kilometers).

Within the core, the increase in pressure leads to transition between the molten outer core and solid inner core, although there may also be chemical differences between the two zones. Recently, seismologists have discovered that the solid inner core does not have the same axis of rotation or the same period of rotation as the rest of Earth – the inner core spins faster. The sharp contrast between the solid inner core and liquid outer core and the fluid motions within the outer core may contribute to the generation of Earth's magnetic field.

The above description suggests that a continuous transition should occur from dense minerals at the base of the mantle to lighter minerals relatively poor in iron and magnesium near the surface of Earth. Accordingly, one would expect a smooth gradation in the composition of Earth.

However, geologists have discovered that while iron-rich and magnesium-rich minerals are indeed the most abundant kinds deep within the mantle, the lighter minerals do not appear gradually, starting at some intermediate depth in the

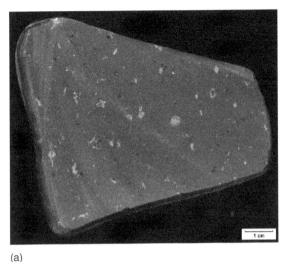

(a)

(b)

mantle, and then steadily increasing in concentration as the surface is approached. Instead, they are confined to a sharply defined layer that rests on top of the denser silicate minerals in the mantle, like a piecrust resting on the filling. This surface layer of lighter minerals resting on the denser minerals of the mantle is called Earth's *crust*. The sharp boundary between the mantle and the crust is called the Mohorovičić discontinuity (after its discoverer Andrija Mohorovičić, a Serbian geophysicist) or Moho for short.

Earth's crust is different under the oceans than on the continents, and this basic dichotomy is a major clue to the geological activity of Earth. Ocean crust is composed of *basalt*, a dark, fine-grained igneous rock made up primarily of the minerals pyroxene and plagioclase feldspar with a little olivine (Figure 8.11(a)). The continental crust is composed largely of *granite*, a light-colored, coarse-grained igneous rock, which is made up of the minerals orthoclase and plagioclase feldspars and about 25% quartz (Figure 8.11(b)). The continental crust does not contain any olivine or pyroxene, and is the least dense part of the solid Earth.

FIGURE 8.11 (a) Basalt, the dark-colored, dense igneous rock that forms the oceanic crust. (b) Granite, the light-colored low-density igneous rock that makes up the continental crust.

USGS

8.7 The floating crust: Isostasy

Measurements of the position of the lower boundary of the Moho show that where the crust is thickest it also extends

FIGURE 8.12 The principle of isostasy, showing how Earth's crust floats on the mantle. Light continental crust floats higher than the denser ocean crust, and the greater the elevation of the continents, the deeper the crustal roots.

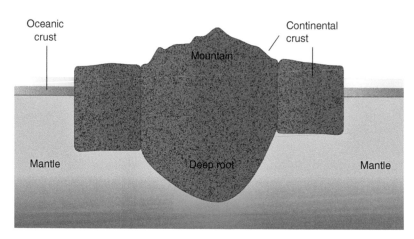

Oceanic crust

Continental crust

Mountain

Mantle

Deep root

Mantle

deepest into the mantle. Apparently the solid rocks of Earth's upper mantle have yielded under the pressure of the thickest, most massive, sections of crust, which have settled into the mantle like blocks of wood settling into water.

One substance floats in another if it has a lower density. For example, wood and ice are less dense than water, and float in water for that reason. It turns out that the rocks of the crust are, in fact, less dense than those of the mantle underneath. Crustal rocks have densities in the range of 2.5 to 2.7 grams per cubic centimeter, compared to a density of about 3.3 grams per cubic centimeter for mantle rock. Rocks that make up the ocean floors are denser (2.7 grams per cubic centimeter) than those of the continental crust (2.5 grams per cubic centimeter). This explains the fact that the crust seems to float in the mantle, with the lighter continental crust floating higher than the denser ocean crust (Figure 8.12).

Only a small part of a floating block of wood or ice rides above the surface of the water; similarly, the visible or "above-ground" part of a large block of crustal material is only a small part of the complete mass of the crust that supports it. The remainder of the mountain block is "submerged" in the mantle. Mount Everest, for example, extends nearly 6 miles (10 kilometers) above sea level, but its continental roots go down about 25 miles (40 kilometers) into the mantle.

Geologists use the term *isostasy* to describe the way in which Earth's crust floats in the mantle. The fact that Earth's mantle can yield under the weight of the crust has important consequences for the geological activity of Earth.

8.7.1 Creep

How can Earth's solid crust "float" in a solid mantle? If something floats in the mantle, that would seem to imply that the mantle is not solid, but liquid. Earthquake waves, by contrast, pass through the mantle in a manner suggesting that the mantle rocks are solid. The answer is a phenomenon called *creep*, which explains how solids can behave like liquids. Creep allows the atoms of a solid to slide over one another, like the atoms of a liquid, when steady pressure is applied to the solid over a long period of time.

To understand why creep occurs in a solid, consider two rows of atoms in a solid (Figure 8.13). The top row contains a hole or defect. A force is applied to the top row from the right (arrow). If this material were a liquid, the atoms in the top row would slide freely over the atoms in the bottom row. That is, a flow would take place. In a solid, the atoms are locked in place by their bonds to their neighbors. However, the atom at the end of the top row is not locked in place as tightly because there is no atom on its left, but only a hole. Therefore, when the force is transmitted down the row of atoms from the right, this atom on the end tends to fall into the hole. In effect, the hole has moved one space to the right.

If the force is applied steadily, the atoms continue to migrate to the left as the hole migrates to the right. At the end, the entire row of atoms has shifted to the left. The two rows of atoms have yielded under the pressure as if they were in a liquid, but much more slowly.

The phenomenon of creep explains the fact that the solid rocks of Earth's interior yield and flow under the weight of mountain ranges on the surface. Although the rocks of the mantle are solid, not liquid, they are quite warm, and, in places, close to their melting point. Consequently, the atoms in these rocks vibrate vigorously, and occasionally break away from their neighbors, leaving holes in the crystal structure. An abundance of such "holes" makes creep possible.

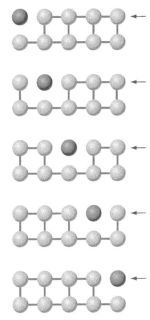

FIGURE 8.13 Illustration of creep, the migration of atoms as a result of holes in the crystal structure of a solid. Through this process, solids, especially when close to their melting temperatures, are able to flow.

Creep in Earth's mantle is a very slow process – too slow to be seen directly. Measurements of the behavior of the mantle under changing forces show that it will flow almost like a liquid when a force is applied for a long period of time.

8.8 Convection currents in Earth's interior

Earth's mantle rests on the iron core, which is very hot, with a temperature of several thousand degrees. The solid rock at the base of the mantle is heated by its contact with the hot core. The heat it absorbs causes the mantle rock to expand. Then it displays its fluidlike properties. Because it has expanded, it is buoyant; like any buoyant fluid (for example, oil in water), it tends to rise to the surface (Figure 8.14). An upward-moving column of warm, solid rock forms and rises slowly toward Earth's surface. The speed of the rising column of rock is only about a tenth of an inch (0.25 centimeters) per year.

Before the upward-moving column of rock actually reaches the surface, however, it runs into a barrier presented by the relatively cold, and therefore more rigid, rocks near the surface. As the materials of the upward moving column approach the surface, they are turned aside by this barrier. Dividing and going in opposite directions, they run horizontally (Figure 8.15).

As the currents of warm rock move horizontally under the surface, they are cooled by their proximity to the cooler surface rocks. When warm rock cools, it becomes more dense. As a result, the currents of rock lose their buoyancy and start to

FIGURE 8.14 Convection in a pot of heated water. When a pot of water is heated on the stove, the water at the bottom of the pot expands and its average density decreases. This heated water rises because of its decreased density. As it rises, the water transfers heat to its cooler surroundings, and in the process cools and becomes denser. It then sinks back to the bottom of the pot to repeat the process. Convection also takes place in the material of Earth's upper mantle, but at a rate millions of times slower.

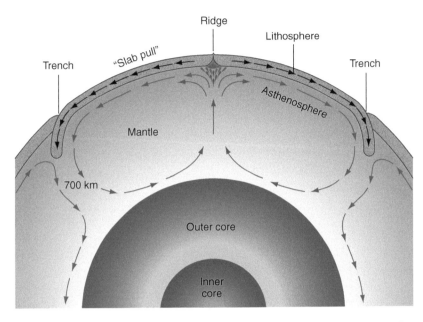

FIGURE 8.15 A rising column of warm, solid rock, deflected by the colder rocks near the surface, divides and runs horizontally at some depth under the surface of Earth. As the moving rock cools, it becomes denser and sinks back towards the core. These convection currents cause the surface layer of Earth to break apart and move horizontally.

sink back into Earth's interior again. When the rock sinks far enough, it is again heated by the core, becomes buoyant, and starts to rise upward once more. The result of these effects is a circulating pattern of movement in Earth's mantle (Figure 8.15).

This pattern of rising and falling motion in a fluid heated from below is called *convection*. The mantle of Earth, or at least a large part of it, may be divided into zones of convection currents. The convection currents in Earth's mantle turn out to be critical to our understanding of the surface features of Earth.

The most striking evidence for convection currents in Earth's mantle is the phenomenon called *continental drift*, which refers to the fact that the continents of Earth are not anchored to Earth's interior, but move or drift across the surface of the planet. Convection in Earth's interior provides an explanation for the movement of the continents. Consider the portion of Earth located above the rising column of rock in Figure 8.15. As noted above, when the column nears the surface, it is deflected by the cold, rigid barrier of the rock near Earth's surface, and divides and flows in opposite directions.

The moving rock tends to drag the overlying material with it. As a result, the surface layer of Earth is pulled in two directions away from the zone directly over the rising column.

The tension of these opposing forces may produce a crack in the surface. The crack widens as the convection currents continue to pull the surface zone apart, and molten material from the upward-moving column (where the rock is closest to melting) wells upward into the crack, nearly filling it. When the molten material congeals, it constitutes a new section of Earth's crust. As time goes on, the two parts of Earth's surface on either side of the crack are dragged farther apart by the underlying convective flows. Where the convection currents turn downward, the overlying material can be dragged down into Earth's interior.

8.9 Earth's lithosphere

How far beneath Earth's surface do the convection currents run? How thick are the portions of Earth's surface that are carried on the backs of these currents? From the description of the way in which the currents arise, it sounds as though they move horizontally directly under Earth's crust, but that turns out not to be so. They must move along at a considerably greater depth than the bottom boundary of the crust. The reason has to do with the details of Earth's internal heating and temperature.

As we noted earlier, heating of Earth's interior, mainly by radioactivity, caused the temperature of Earth to rise until it approached the melting point of rock. Today, however, only a small part of Earth's mantle is actually close to melting. In fact, the melting and near-melting is limited to a narrow zone, about 60 miles (100 kilometers) below the surface, and even in this narrow zone, melting is only partial.

The explanation for the narrowness of the zone of partial melting can be found in a comparison of the temperature at various depths in Earth with the temperature at which rocks melt. A graph of Earth's internal temperature (Figure 8.16) shows that Earth's temperature rises rapidly in the upper mantle to about 1500 °C at a depth of about 60 miles and then rises more slowly in the lower mantle. Eventually the temperature reaches a value of roughly 5000 °C at the center of the planet.

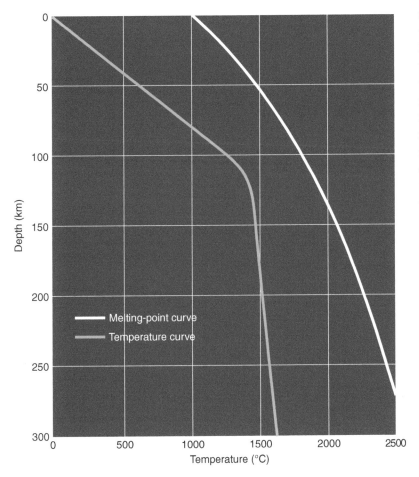

FIGURE 8.16 Graph of temperature vs. depth in Earth's mantle also showing the melting point of rock at various depths. Earth's internal temperature rises rapidly in the uppermost mantle and then much less rapidly below about 60 miles (100 kilometers). The melting point temperature of rock increases with depth because the increasing confining pressure makes melting more difficult. At depths of about 60 miles in the upper mantle the two curves approach one another and the rocks at those depths are close to their melting point.

The figure also shows the temperature at which rocks melt in Earth's mantle. The melting-point temperature curve has two important characteristics:

First, the melting-point temperature of rock is not the same everywhere within Earth. It increases with increasing depth, because at great depths the pressure locks the atoms of the solid rock into place. Therefore, deep within Earth a higher temperature is required to break the bonds that tie these atoms to their neighbors, and convert the solid rock to a liquid. For that reason, the rocks at great depth in Earth's mantle are solid, and not melted, even though they are very hot.

Second, the melting-point temperature of rock is considerably higher than Earth's internal temperatures near the surface. The reason, of course, is that Earth loses its heat to space at the surface and is relatively cold there. Consequently, the rocky material near the surface is also solid.

In the narrow zone about 60 miles (100 kilometers) below the surface, the rocks can be partly molten, or so close to melting that, while solid, they are relatively soft and yielding. For that reason, the region at that depth is called the *zone of weakness* (also called the *asthenosphere*).

The layer of rock above the zone of weakness, as noted above, is relatively cold and rigid. This rigid layer of rock, about 60 miles thick is called the *lithosphere*. The lithosphere is much thicker than Earth's crust, which has an average thickness of about 8 miles (13 kilometers). However, it is still quite thin in comparison with the radius of Earth. The lithosphere is best thought of as a bit of relatively rigid material capping Earth, with the continents and other materials of the crust embedded in it like logs frozen into an ice floe. These sections of Earth's surface average thousands of miles or kilometers across, or more than 20 times their thickness. In fact, in proportion to their extent, the moving pieces of lithosphere have about the same thickness as an eggshell.

Because these sections of lithosphere are so thin compared to their extent, geologists call them *plates*. According to the available evidence, the entire surface of Earth is divided into about a dozen such plates. The movement of Earth's plates has created the major surface features of Earth, from continents and ocean basins to mountain ranges and volcanoes.

8.10 Summary

By the end of this chapter you should be familiar with the following concepts and topics:

Structure and composition of Earth
Composition of Earth
The chemical bond
Formation of rocks
  Properties of rocks

Questions

8.1. What are the eight most abundant elements on Earth? Explain why these elements are abundant in terms of: (1) stellar evolution, and (2) the origin of the inner planets of the Solar System.

8.2. What were the major sources of heat for the melting of the early Earth?

8.3. What is the composition of Earth's core? How do we know this? What is the composition of Earth's mantle and crust?

8.4. Why does core formation indicate that the early Earth was molten or largely molten?

8.5. Explain why the Moho is located at a greater depth under high mountain ranges than under the deep ocean basins.

8.6. What is creep?

8.7. Where do convection currents originate within Earth? What effect do these convection currents have on the lithosphere of Earth?

CHAPTER 9

The changing face of an active habitable planet

Explanation of Earth's surface features

The evolution and distribution of life on Earth is linked to the changing face of our planet. The idea that the continents are not fixed in place, but can drift across Earth's surface seems as strange now as it did when first proposed. Since the 1960s, much geological and geophysical information has accumulated to provide clear evidence that the continents move. The theory of plate tectonics explains the major features of Earth, and the history of Earth's crust over billions of years. As plates separate, move, and collide, they also affect the oceans and atmosphere. On timescales of millions of years, plate movement is a major cause of climate change, which in turn has greatly influenced the history of life on Earth.

Earth's continents were once a single block of land called Pangaea, which subsequently broke apart into the continents we see today. In this chapter we will learn what happens when continental plates separate and collide, and see that most of the Earth's major surface features can be explained by the theory of plate tectonics, from mountain ranges to volcanoes, narrow oceans, and rift valleys. We will follow the trail of evidence for this theory, from the distribution of fossil plants and animals across continents to the record of changes in the Earth's magnetic field. We will see how the motion of the continents can be directly measured, and how the plate boundaries have been located by pinpointing areas of earthquake and volcanic activity.

9.1 Continents and ocean basins

The continents are the most elevated regions of Earth's crust because they are composed largely of the low-density igneous rock *granite*, which floats relatively high on Earth's mantle. The ocean floors are composed of the dense igneous rock *basalt*, which rides lower on the mantle. Hence, the ocean basins are the lowest regions of the planet's surface.

The earliest continents began to form at least 4.3 billion years ago. The first light crust was most likely produced during the melting of Earth caused by the intense early bombardment of the planet. As discussed earlier, molten iron sank to the planet's center forming the core, and the lightest minerals rose to the surface to form the initial continental crust. These primordial continents were apparently much smaller than the present continents. The area of continental crust grew over time through the rise of additional granitic magma to Earth's surface in areas around the ancient cores of the continents.

For many years, geologists believed that the continents and ocean basins were fixed in place, and had not changed in relative positions since they were first created. By the early twentieth century, however, some geologists began to question this assumption, and proposed that the continents have moved great distances horizontally.

Previous page: The eruption of Mount St. Helens, May 18, 1980.
Photo by Austin Post, Skamania County, Washington, Courtesy of USGS

9.2 Evidence for continental drift and plate tectonics

The evidence for continental drift was first incorporated into a comprehensive theory by the German meteorologist Alfred Wegener (1880–1930) in 1912 (Figure 9.1). Wegener had noticed that the western coastline of Africa fit surprisingly well into the eastern coastline of South America. He proposed that the fit of the continents around the Atlantic Ocean is so good, especially if the edges of the continental shelves (the true edge of the continents) are used as the basis for the fit, that the only reasonable conclusion is that the two continents were joined together in the past and later broken apart (Figure 9.2).

Wegener presented evidence that all of the present continents were together in one unit block about 250 million years ago. He called the former super-continent *Pangaea*, meaning "all lands." According to Wegener, Pangaea broke apart and the pieces moved to their current positions to become the present continents. As the continental blocks moved apart,

FIGURE 9.1 Alfred Wegener (1880–1930), early proponent of continental drift. Wegener perished during an expedition to Greenland in which he was attempting to measure continental movements directly.

© Alfred Wegener Institute for Polar and Marine Research, Bremerhaven, Germany.

sliding on a warm fluid layer in the upper mantle, the ocean floor was produced by new material rising up from this partially molten layer.

Wegener's ideas met considerable resistance from many geologists who were committed to a picture of Earth where continents and ocean basins had remained in fixed positions throughout geological time. Despite considerable evidence favoring the movement of continents over time, the theory of continental drift was rejected by the geological community. New evidence that came to light in the 1950s and 1960s led to a revolution in the earth sciences in which continental drift was shown to be a reality.

9.2.1 Match of fossil flora and fauna

Another line of evidence for continental drift first recognized by Wegener comes from the distribution of fossil plants and animals. The fossil record shows that identical species of animals and plants are found on continents now separated

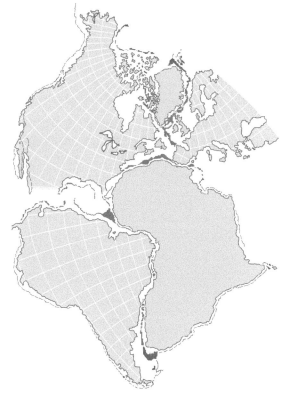

FIGURE 9.2 The fit of the continents around the present Atlantic Ocean.
USGS

FIGURE 9.3 Evidence for continental drift from the distribution of ancient animals and plants. The distribution of the fossil reptile *Mesosaurus* in South America and South Africa about 270 million years ago is shown in blue. The distribution of the 250-million-year-old fossil plant *Glossopteris* is shown in green. The distribution of other fossil animals that support continental drift is also shown.

USGS

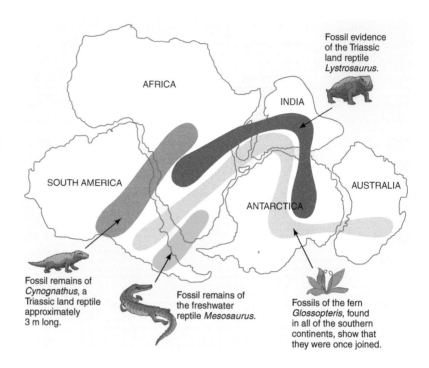

by wide oceans. For example, remains of *Mesosaurus*, a reptile that lived about 270 million years ago, are found on both sides of the Atlantic, in South America and Africa (Figure 9.3). This reptile's range covers regions that would be contiguous if the two continents were fitted together.

Mesosaurus could not have traveled across the wide ocean that now separates the two continents, and it is very unlikely that the same species would have evolved independently on two separate continents. The most straightforward explanation is that *Mesosaurus* lived at a time when Africa and South America were joined together.

Another example is the 250-million-year-old fossil plant *Glossopteris*, which is found on all the present southern continents, as well as on the Indian subcontinent. The distribution of *Glossopteris* fossils supports the reconstruction of these continents as one large continental block covered by forests containing *Glossopteris* (Figure 9.3).

9.2.2 Match of rock units

Further evidence comes from rocks of matching types and ages found in areas of South America and Africa that would be

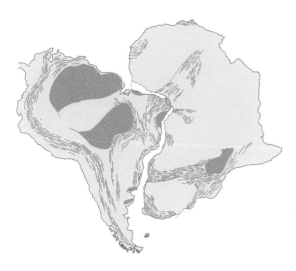

FIGURE 9.4 Fit of West Africa against South America showing contiguous regions of rocks of similar ages. The dark areas represent ancient rocks more than 1900 million years old. The lighter areas are younger rocks formed less than 600 million years ago.

contiguous if the two continents were fitted together. For example, areas with rocks younger than 600 million years old in Africa abut sharply against adjacent regions of older rocks that were formed some 1900 million years ago. Rocks of the same ages, with the same sharp boundaries are found in South America. When the two continents are brought back together along their matching edges, the rock types and their boundaries line up almost perfectly (Figure 9.4).

9.2.3 Evidence from the ocean floor: The Mid-Ocean Ridge

Little was known about the ocean floor when Wegener proposed his theory. In the 1950s, geologists began to explore the ocean basins with new instruments capable of recording the features of the ocean floor in some detail. Relief maps of the Atlantic Ocean floor constructed from this information showed a bulge or elevation roughly equidistant between the continental shorelines on either side (Figure 9.5). This elevated region is called the *Mid-Ocean Ridge*. The Mid-Ocean Ridge occurs in all the ocean basins as a continuous feature some 40 000 miles (64 000 kilometers) long. It winds around Earth much like the seams on a baseball (Figure 9.6).

Running along the center of the ridge is a crack called the *rift valley*. This central crack, discovered by oceanographers in the late 1950s, is very active geologically. The rift valley is marked by abundant earthquakes, undersea volcanism, and the escape of a large amount of heat from Earth's interior.

FIGURE 9.5 A relief map of the Atlantic Ocean floor showing the Mid-Ocean Ridge. The crack, or rift valley, running along the center-line of the ridge is evident.
   NOAA

FIGURE 9.6 The Mid-Ocean Ridge system – the globe-encircling crack in the ocean floor is shown by the light-blue linear feature that runs through all of the ocean basins.
   NOAA

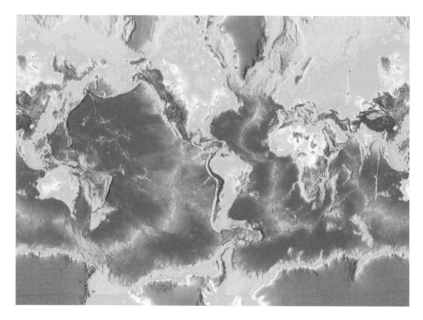

We now know that this great crack in the crust marks the place where two sections of Earth's crust are moving apart. As the blocks of crust move apart, hot rock in the crack between them is exposed to the lower pressure of the surface. Since rocks melt more readily at low pressure, the decrease in pressure in the crack causes some of the rock at depth to melt. The molten rock rises, and the magma fills the crack. In time, the molten rock cools and solidifies, forming a new addition to the ocean floor. This process, which works like a conveyer belt, is called *sea-floor spreading*; it was first proposed by the geologists Harry Hess and Robert Dietz in the early 1960s.

The parallel cracks in the Mid-Ocean Ridge running at right angles to the central rift seen in Figure 9.5 are called *fracture zones*. These great faults break the Mid-Ocean Ridge into numerous segments. The discovery of these features provided further support for sea-floor spreading and plate tectonics. The segmentation of the ridge is necessitated by the fact that movement of the rigid ocean floor is occurring on a spherical Earth, and some sideways motion must be accommodated as the ocean grows by sea-floor spreading. Only those sections of the fracture zones, called *transform faults*, between the offset rift valleys show active sideways motion and earthquake activity.

9.2.4 Evidence from Earth's magnetism

The evidence that convinced most geologists of the reality of sea-floor spreading, and hence continental drift, is the record of changes in Earth's magnetic field. To understand this critical evidence, we must first discuss the nature of Earth's magnetism and its origins.

Earth behaves as if a giant bar magnet sits at the center of the planet, aligned roughly with Earth's axis of rotation. This internal magnet generates a global magnetic field with north and south magnetic poles close to Earth's geographic North and South Poles (Figure 9.7).

Earth's magnetic field is produced by movements of liquid iron–nickel in the outer core. The combination of convection in the outer core and Earth's rotation causes flow in the liquid outer core to align with the Earth's axis of rotation. Earth's magnetism seems to be "self exciting;" that is, the moving iron–nickel in the outer core creates an electric current that generates a magnetic field. The iron–nickel moving within the magnetic field then creates an electric current, which in turn generates a magnetic field, and so on.

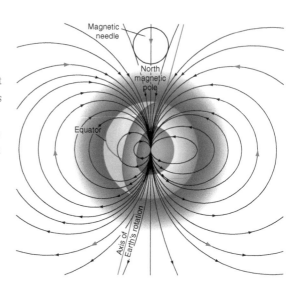

FIGURE 9.7 Earth's magnetic field. Earth behaves as if it has a giant bar magnet at its center, with the magnet's poles close to the rotational axis of the planet. A magnetic compass needle aligns with the north–south field. If the needle is free to rotate, it will line up with the magnetic field lines such that the needle is horizontal at the Equator and vertical at the magnetic poles.

© 2000 John Wiley & Sons, Inc. New York. Skinner, B. J. & Porter, S. C., *The Dynamic Earth* (4/e) Figure 8.16. This material is used by permission of John Wiley & Sons, Inc.

One of the most remarkable geological discoveries of the twentieth century was the realization that Earth's magnetic field actually reverses from time to time. Scientists working on basaltic lava flows on volcanic islands like Hawaii found that these rocks faithfully recorded the direction of Earth's magnetic field at the time they formed. As the basaltic lava cools, small crystals of the iron-rich mineral magnetite (Fe_3O_4) form and orient themselves, like tiny compass needles, in the direction of Earth's magnetic field.

It was found that in a series of ancient lava flows stacked one on the other, some flows recorded the direction of Earth's magnetic field as it is today, whereas other flows recorded the opposite magnetic field direction (i.e., the north magnetic pole was at the south geographic pole) (Figure 9.8). In time, geologists worked out a sequence of about 20 flips of the magnetic field going back more than 4 million years (Figure 9.9). Eventually, a complete magnetic timescale was created, with more than 100 reversals of Earth's magnetic field going back some 200 million years.

9.2.5 Magnetic stripes on the ocean floor

The strength of Earth's magnetic field can be calculated for any place on Earth, taking into account that the intensity of the field varies with geographic location in a predictable way. The calculation presupposes that the near-surface

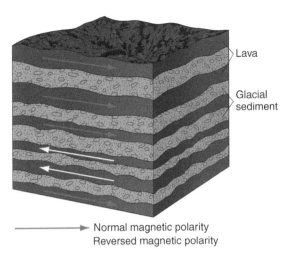

Normal magnetic polarity
Reversed magnetic polarity

FIGURE 9.8 A series of lava flows (separated by sediment layers) in which a reversal of Earth's magnetic field is recorded.

ⓒ 2000 John Wiley & Sons, Inc. New York. Skinner, B. J. & Porter, S. C., *The Dynamic Earth* (4/e) Figure 8.16. This material is used by permission of John Wiley & Sons, Inc.

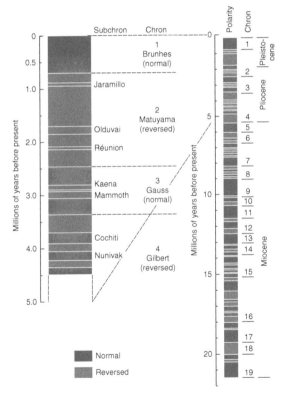

FIGURE 9.9 Record of reversals of Earth's magnetic field going back more than 20 million years (at right). The record of the last 4.5 million years (at left) comes from radiometric dating of lava flows. Major intervals of a given polarity are called Polarity Chrons and shorter intervals are labeled Subchrons.

ⓒ 2000 John Wiley & Sons, Inc. New York. Skinner, B. J. & Porter, S. C., *The Dynamic Earth* (4/e) Figure 8.16. This material is used by permission of John Wiley & Sons, Inc.

rocks do not contain unusual quantities of magnetic minerals.

By contrast, a magnetometer towed behind a ship measures precisely the actual intensity of Earth's magnetic field as the ship moves along. If the measured and calculated values of intensities for any specific location do not agree, a *magnetic anomaly* is said to exist. If the measured intensity is greater than the calculated value, the magnetic anomaly is said to be positive. If the measured intensity is smaller than the calculated value, the magnetic anomaly is said to be negative.

In the 1960s, oceanographers surveying the magnetic field of Earth over the oceans discovered magnetic anomalies in the form of areas of enhanced or reduced magnetic field strength that formed a pattern of stripes running parallel to the center-line of the Mid-Ocean Ridge.

In 1963, geophysicists Fred Vine and Drummond Matthews of Cambridge University realized that a combination of sea-floor spreading and reversals of Earth's magnetic field would produce just this kind of pattern.

During the process of sea-floor spreading, new ocean basins open by spreading outward from the center of the Mid-Ocean Ridge in a conveyor-belt fashion. When two sections of Earth's surface are separating, molten material comes up into the crack or rift between the two sections of crust. As the molten rock approaches the surface, it cools and solidifies. The ocean-floor rocks contain a small amount of iron-rich minerals that becomes magnetized in the direction of Earth's magnetic field when the rock solidifies.

If the two sections of Earth's surface are moving apart at a typical rate of about 2 inches (5 centimeters) per year, then after 500 000 years the ocean plates have moved apart about 16 miles (25 kilometers). At the end of that time, there is a parallel band or "stripe" running down either side of the crack or rift, in which the rocks are magnetized in the present (or normal) direction of Earth's magnetic field.

But Earth's magnetic field flips or reverses direction roughly every half million years. Suppose one of these reversals occurs at this junction. The magnetic field now points in the opposite direction (i.e., south), and fresh molten material, rising up as the plates continue to separate, will be magnetized in this reversed direction as it cools and solidifies.

Now we have two stripes running parallel to each other on either side of the rift. One is magnetized normally and the other magnetized in a reverse direction. As Earth's magnetic

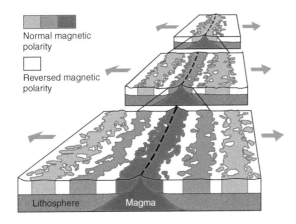

Normal magnetic polarity

Reversed magnetic polarity

Lithosphere   Magma

FIGURE 9.10 The development of the ocean floor through time by sea-floor spreading as evidenced by the magnetic stripes.

field flips back and forth, the ocean floor will acquire a pattern of symmetrical magnetic stripes with alternating directions of magnetization (Figure 9.10).

Measurements of the magnetization of rocks from all the ocean floors show just this alternating pattern of symmetrical magnetic stripes. It is very difficult to explain the pattern in any way except as the result of molten rock coming up into the crack between two separating sections of Earth's crust and solidifying there, recording the direction of Earth's magnetic field at the time it solidified.

The width of each stripe depends on the rate at which the two parts of Earth's surface were moving apart when that stripe was formed, and the length of time the magnetic field retained that direction before flipping. Since the latter is known from magnetic studies of lava flows and other rocks, the former can be inferred. Thus, the magnetic stripes can tell us the speed at which the plates have been moving apart for millions of years in the past.

Combined with other evidence, the magnetic stripes enable us to reconstruct the positions of the continents in the past. They tell us that at one time in the past, all the continents were assembled, as a single super-continent, called *Pangaea*. It also tells us when and how Pangaea began to break up into separate pieces – the Americas, Africa, Eurasia, and so on. The observation of these stripes of magnetization in the ocean floor is strong confirmation of the validity of the theory of plate movements and drifting continents.

9.2.6 Age of ocean-floor rocks

The age of the ocean floor can be predicted by estimating the rate of sea-floor spreading from the distances of magnetic stripes of known age from the center of the Mid-Ocean Ridge. The ages of the rocks of the ocean crust have also been determined directly by drilling down to bedrock in the ocean basins. Results show that the youngest rocks on the ocean floor are at or near the rift (Figure 9.11). Furthermore, the ocean-floor rocks get progressively older with distance from the central rift. This pattern matches that predicted by the pattern of magnetic stripes and supports the picture that the rocks in and near the rift have recently risen to the surface as molten material from the mantle beneath, into the space between the separating plates.

The oldest rocks in the Atlantic Ocean, encountered at the edges of the continents that border the ocean on either side, have ages of about 185 million years. These are the rocks that must have been formed when the original continent first broke in two, and the two pieces began to separate. Apparently, prior to 185 million years ago there was one continent in this region, and no Atlantic Ocean. The present Atlantic began to form when the break occurred in the ancestral continent, and has been widening ever since that time, as the continents on its eastern and western shores continue to move apart.

9.2.7 Direct measurements of plate motions

The motion of the plates can be measured directly by using satellites that orbit Earth at the same rate as the planet's rotation, and thus remain in fixed positions in space relative to Earth. The Laser Geodynamics Satellite (LAGEOS) provides a stationary target for laser beams directed upward from stations on the ground. Measurements of the time required for the beam to travel from the ground to the satellite and back allow scientists to calculate the exact distance between the ground station and the LAGEOS satellite. Measurements made from year to year from various ground stations have shown that the distances between the stations and the satellite have undergone systemic changes, and thus that the continental landmasses are moving.

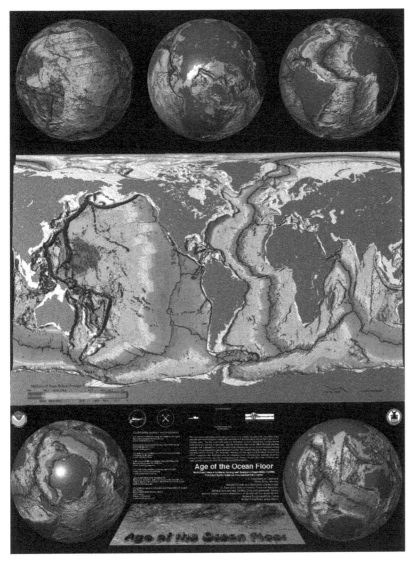

FIGURE 9.11 The age of the crust in the oceans. The age of the ocean floor is shown as bands of different color on the basis of the magnetic striping developed during sea-floor spreading. The youngest ocean floor is near the Mid-Ocean Ridge, while the oldest is furthest away.
    NOAA

The Global Positioning System of satellites sends radio signals to receivers on the ground so that the location of any point on Earth can be determined to within a few millimeters. Thus, geologists can use these satellites to measure the movements of the plates in real time. The results of the satellite measurements are consistent, and show that the slowest plates travel 0.4 to 1.2 inches (about 1 to 3 centimeters) per year, whereas the fastest plates move up to 4 inches (10 centimeters) per year. These rates agree with rates of plate motion estimated from geological evidence, such as the widths of the magnetic stripes on the ocean floor.

9.3 Boundaries of Earth's plates

The picture that has emerged is of the entire surface of Earth broken into a mosaic of plates, which move across the surface, colliding with neighboring plates in some cases and separating from them in others. In a number of places, one plate grinds past the adjoining plate. Each of these types of relative motion at a boundary between two plates can be expected to

FIGURE 9.12 The global distribution of earthquakes. The boundaries of Earth's major plates are the scenes of violent seismic and geological activity. Most earthquakes occur where plates separate, collide, or slide past one another.
   NOAA

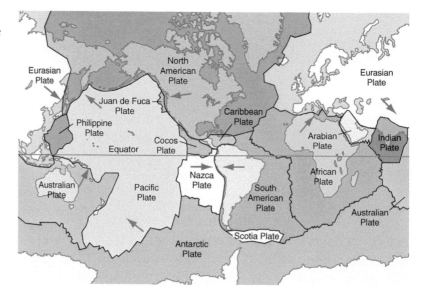

FIGURE 9.13 Map of Earth's major plates. This map of the plates also shows the approximate direction (arrows) of motion of each major plate.
   USGS

cause earthquake disturbances on the surface and in the interior of Earth. This suggests that if we want to locate the boundaries of Earth's plates, we should look for regions in which there seems to be an exceptional concentration of earthquakes.

A map of all earthquakes recorded on Earth for several years reveals that most earthquakes are concentrated in narrow zones (Figure 9.12). Before the advent of plate tectonic theory, geologists could not explain why the world's earthquakes tended to occur in such narrow regions. Now we can see that these zones can be used to define the boundaries of Earth's plates.

The occurrence of other kinds of geological activity, especially volcanism, along these same narrow zones lends support to the idea that these sites represent plate boundaries. The result is a map of Earth's plates (Figure 9.13). The surface of Earth is made up of about a dozen large plates and several smaller plates. Some of the plates are entirely oceanic (e.g., Pacific Plate, Nazca Plate, and Cocos Plate), and some are a combination of oceanic and continental areas (e.g., North and South American Plates, African Plate, and Eurasian Plate).

9.4 Types of plate boundaries

9.4.1 Separating or diverging plates

The boundary between separating plates is marked by the Mid-Ocean Ridge. As described above, the Mid-Ocean Ridge contains a central rift or crack, marked by active volcanism. As the plates separate, magma rises to fill the crack and congeals. As the plates move further apart, new magma is emplaced in the growing crack. In this way, new ocean floor is produced as the ocean plates move apart (Figure 9.14).

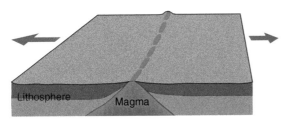

FIGURE 9.14 A divergent plate boundary – a Mid-Ocean Ridge.

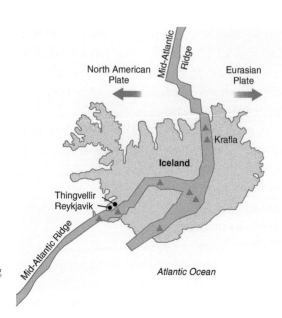

FIGURE 9.15 Map of Iceland, showing its position along the crest of the Mid-Atlantic Ridge.

USGS

If the cracking or rifting takes place within a continent, basalt magma also rises from Earth's mantle to fill the crack. As the two continent pieces move apart, continued basaltic volcanism creates new sea floor between the two continental fragments. Eventually, a true Mid-Ocean Ridge system develops on the newly formed ocean floor.

The best place in the world to see this process in action is Iceland, a volcanic island that straddles the Mid-Atlantic Ridge (Figure 9.15). The central rift of the ridge runs across Iceland splitting the country in two (Figure 9.16). The rift is marked by wide fissures, volcanic activity, and earthquakes.

9.4.2 The opening of a new sea

This process of new ocean formation is taking place today in the Middle East. A continental plate is breaking up, and a new sea is opening. As Figure 9.17 shows, Egypt and the Sudan in Africa are divided from Saudi Arabia by the Red Sea. Examination of the rocks on the floor of the Red Sea shows that they are freshly minted ocean crust, the kind of material that wells up as lava at the Mid-Ocean Ridge, and then congeals to form a new addition to the ocean floor. The Red Sea is also a region of earthquakes and exceptionally high heat flow from Earth's interior. These findings indicate that the Red Sea

FIGURE 9.16 Thingvellir, Iceland, where the rift valley of the Mid-Atlantic Ridge can be seen cutting across the island.

Michael Rampino

is a new ocean just beginning to open up, just as the Atlantic started to open about 185 million years ago. According to measurements on the ages of rocks on the floor of the Red Sea, it began to spread apart about 10 million years ago.

At its southern end, the Red Sea meets the African Rift Valleys, which run down through East Africa. The Rift Valleys are a region of Africa in which the continent was uplifted and then fractured. The deep fractures have been partially filled by sediments to form a series of broad valleys. Earthquakes and volcanic activity occur along the Rift Valleys, further suggesting that they represent a region of incipient plate breakup. If the present activity continues, this entire region may split off from the rest of Africa to become a small separate continent.

9.4.3 Colliding plates: Ocean against ocean

When two ocean plates are pushed together, one of the plates is forced beneath the other. This process is called *subduction*, and the region where this takes place is called a *subduction zone*. As one plate descends beneath the other, a great geological upheaval takes place. The down-going slab of ocean

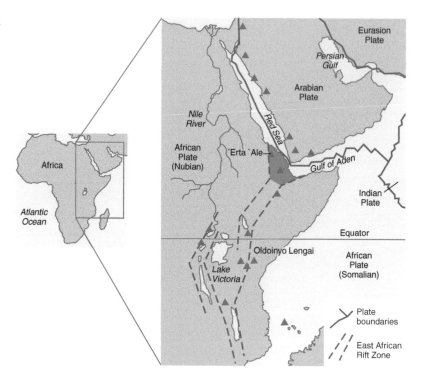

FIGURE 9.17 The Red Sea and East African Rift Valleys.
    USGS

plate is bent downward creating an especially deep area of the oceans called an *ocean trench*. The heat generated by the friction between the two grinding plates causes melting of the down-going slab. The resulting magma rises to the surface through cracks in the upper plate to form a curved chain of volcanic islands, a so-called *volcanic island arc*.

The grinding action between the two plates also causes earthquakes that occur from Earth's surface down to a depth of more than 400 miles (640 kilometers) along the down-going plate. Subduction zones are marked by deep trenches in the ocean floor, volcanic island arcs, and steeply dipping zones of deep earthquakes (Figure 9.18). The Aleutian Islands off the coast of Alaska and the Philippines in the western Pacific are good examples of this kind of subduction zone.

9.4.4 Colliding plates: Ocean against continent

If an ocean plate is pushed against a plate that has a continent on it, the denser ocean plate is always subducted beneath the lighter continent. A deep ocean trench is formed by the down-bending of the ocean plate, and again there is friction between

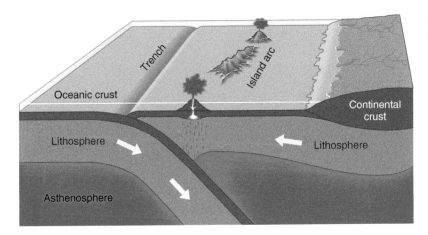

FIGURE 9.18 A subduction zone in which one ocean plate dives beneath another ocean plate.
    USGS

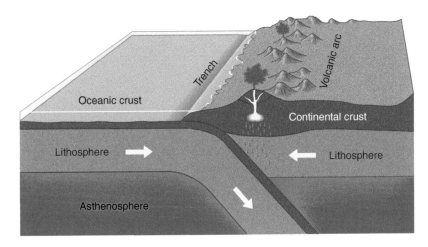

FIGURE 9.19 A subduction zone where an ocean plate dives beneath a plate containing a continent, producing a mountain range like the Andes.

the descending slab and the upper plate that causes melting of the down-going plate. Magma then rises to the surface through fractures in the continent, and forms a chain of volcanoes somewhat inland from the coast (Figure 9.19). The Cascade volcanic chain, including Mount St. Helens and Mount Rainier, in the northwestern United States is an example of this kind of volcanic activity. In this region, the small oceanic Juan de Fuca Plate is being subducted under the North American Plate (Figure 9.20). The magma feeding these volcanoes originates from partial melting of a mixture of the basaltic ocean crust and the overlying sediment, and when solidified produces a rock intermediate in silica content between a basalt and a granite. This kind of rock is called *andesite*, after the Andes Mountains of South America, where

FIGURE 9.20 Subduction of the oceanic Juan de Fuca Plate beneath the North American Plate produces the volcanism of the Cascade Range.
USGS

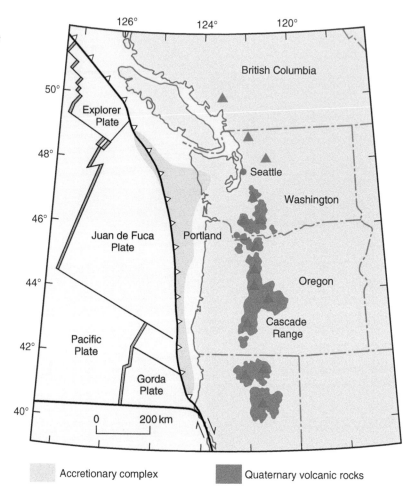

the oceanic Nazca Plate is being subducted beneath the South American Plate.

Eventually the heating at the base of the continent causes the plate to bulge upward in a great uplift, and the compression caused by the two plates pushing against each other crumples the rocks at the edge of the continent into great folds. These uplifted and crumpled rocks form a chain of mountains.

The Andes are a good example of a mountain range formed by a combination of deformation, uplift, and volcanism, as the Nazca Plate is subducted under the western edge of the South American continent. The highest peaks in the Andes are massive volcanoes, sitting on top of the uplifted and deformed western edge of South America (Figure 9.21).

FIGURE 9.21 The Andes Mountains of western South America. In the foreground, upturned layers of sedimentary rocks that were once horizontal can be seen.
  George Ericksen, USGS

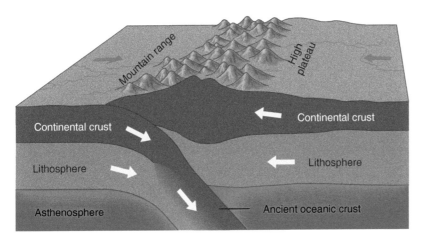

FIGURE 9.22 Collision of two continental plates forming a mountain range like the Himalayas.

9.4.5 Colliding plates: Continent against continent

Ocean plates can be completely subducted, bringing into contact the two continents that were at opposite edges of the ocean basin. Continents are built of lighter, more buoyant materials than the oceanic plates. Therefore, when two continents meet neither can be subducted, and the two continents collide with dramatic consequences. Like the fenders of two automobiles crashing together, the edges of the two continents become crumpled and uplifted into a great mountain range (Figure 9.22).

A good example of a collisional mountain range is the Himalayas that lie between Asia and the Indian subcontinent. At the time of Pangaea, India was located far to the south of

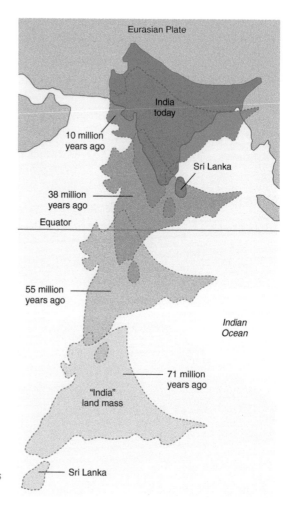

FIGURE 9.23 The movement of India northwards resulted in the eventual collision of two continental plates, India and Asia, creating the Himalaya Mountains and the Tibetan Plateau.

the Equator, and attached to the Antarctic continent. About 130 million years ago, the plate containing India broke free of Antarctica and started to move due north toward Asia at relatively high velocities, at times as great as 6 inches (15 centimeters) per year. Around 40 million years ago (Figure 9.23), India crossed the Equator in its northward journey and approached Asia.

The collision between the Indian and Asian continental landmasses over the last 40 million years has deformed and uplifted the crust at the boundary between northern India and southern Asia to form the Himalayas. Sediments that had lain at the bottom of the sea that once separated India from Asia were uplifted and became the tops of mountains; when

Edmund Hillary and Tenzing Norgay climbed Mount Everest, they found fossilized sea shells near the summit.

The northward movement of India continues today, but at a considerably reduced speed. The result is that India has been pushed part way under Asia so that the continental crust of central Asia has doubled its usual 15-mile (24-kilometer) thickness beneath the Himalayas and the Tibetan Plateau. This extra thickness of buoyant continental crust accounts for the unusually high mountains and plateau area – the highest region on Earth.

9.4.6 Sliding plate boundary

The final kind of plate boundary occurs when two plates move past each other along a great break or *fault* in Earth's crust. The best-known example of this type of boundary is found in California, where the Pacific Plate is sliding past the North American Plate along the San Andreas Fault (Figure 9.24).

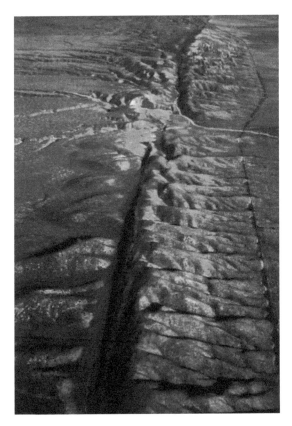

FIGURE 9.24 Sideways plate motion of the Pacific Plate past the North American Plate produces the San Andreas Fault, clearly visible in California.

Robert E. Wallace, USGS

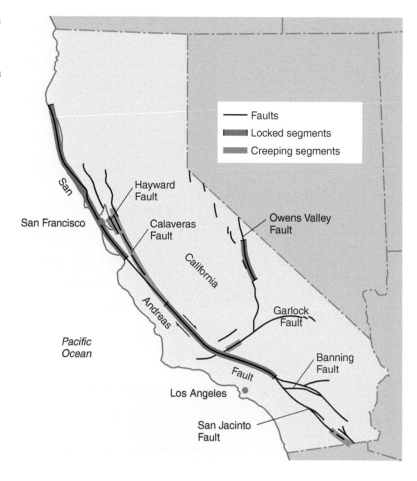

FIGURE 9.25 The San Andreas Fault forms the boundary between the North American and Pacific Plates in California along the west coast of the United States.

USGS

This fault is marked by large numbers of small to moderate sized earthquakes, and occasional great earthquakes such as the 1906 San Francisco event (Figure 9.25). The largest earthquakes occur along sections of the fault line that become locked, and move only after 100 years or so of stress has built up along the fault. The sudden large shift and the release of energy is catastrophic, causing violent movements of the ground in areas close to the fault. The 1906 earthquake and resulting fire destroyed much of the city of San Francisco.

9.4.7 The Hawaiian Islands and hotspots in Earth's mantle

According to the explanation of Earth's moving plates, earthquakes and volcanoes generally occur near the boundaries of plates. But what are we to make of the Hawaiian Islands, a

FIGURE 9.26 A volcanic eruption on the island of Hawaii. Hawaii, one of the most active volcanic islands in the world, sits in the middle of the Pacific Plate, far from any plate boundary. USGS

cluster of active and extinct volcanic islands almost exactly in the middle of the Pacific Plate, far from any plate boundary? The Hawaiian Islands seem to violate one of the principal rules of plate tectonics (Figure 9.26).

FIGURE 9.27 The Hawaiian Island Chain. These volcanic islands are created as the Pacific Plate moves over a fixed hotspot in Earth's mantle.

USGS

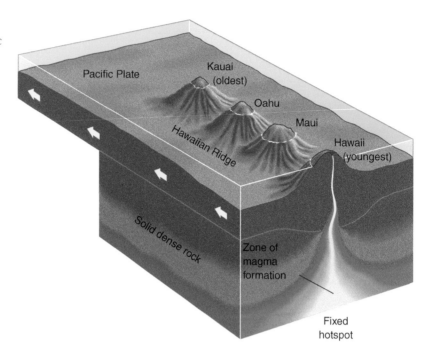

The answer is that Hawaii lies over a *hotspot* – a place where an upwelling column of hot material from deep within Earth's mantle rises to the surface. According to the geological evidence, this hotspot has been erupting molten rock, on and off, for more than 70 million years. As the series of eruptions took place, the Pacific Plate moved, more or less steadily, over the hotspot. Each eruption episode deposited a pile of congealed lava on the surface, making a volcanic mountain, or more accurately, a volcanic island, since the volcano was forming in the middle of the ocean.

The Pacific Plate continued to move, carrying the pile of congealed lava with it. Somewhat later, another series of eruptions occurred, and another pile of lava accumulated on the surface, forming another volcanic island. One by one, the new volcanic islands formed over the hotspot, and then were carried away by the moving Pacific Plate, as if they were on a conveyor belt (Figure 9.27).

A remarkable finding supports this picture. A chain of islands and submerged volcanoes, or seamounts, stretches across the Pacific in a northwest direction from Hawaii. One of the islands in the chain is Midway Island, site of a famous naval battle of the Second World War. The remarkable aspect of this island chain is the fact that the ages of the rocks in these volcanic islands and

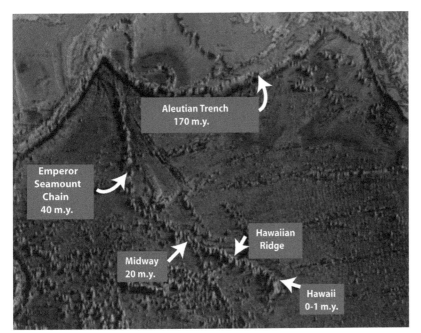

FIGURE 9.28 Hotspot volcanism and the Hawaiian Islands. For at least 70 million years, the Hawaiian hotspot has been essentially stationary, while the Pacific Plate has moved to the northwest some 4000 miles (6400 kilometers). The oldest volcanoes, at the northwest end of the chain, have been so eroded that they no longer rise above the surface of the ocean. The hotspot is now building a new volcano, called Loihi, to the southeast of the island of Hawaii.
USGS

seamounts have been measured, and show an orderly progression. The youngest rocks occur in the present-day Hawaiian Islands, and the islands and seamounts get progressively older as one moves along the chain to the northwest (Figure 9.28).

This line of extinct volcanoes is exactly what one would expect from the ocean floor moving over the hotspot. The islands currently over the hotspot, namely Hawaii, have just been made, and are the youngest. If the Pacific Plate has been moving over the hotspot at a roughly constant speed, the other islands in the chain should have ages proportional to their distance from Hawaii. This is just what we see in Figure 9.28.

The chain of islands and seamounts also shows a distinct bend or change in direction, to an almost due north direction, for seamounts 40 million years old or older. The oldest seamount in the chain is about 70 million years old, and is thousands of kilometers to the north and west of Hawaii, near the Aleutian Islands.

What is the significance of the bend in the chain of volcanoes older than 40 million years? Apparently the Pacific Plate was moving almost due north prior to 40 million years ago, but then its direction of motion changed, and since that time the plate has been moving to the northwest.

FIGURE 9.29 Earth's major hotspots are shown as red dots. Several of the hotspots are labeled. The Mid-Ocean Ridge is shown as a red line, and subduction zones are represented by blue lines with teeth. Shaded areas are plate boundary zones.

USGS

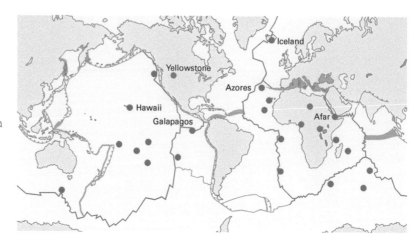

Hawaii is not the only hotspot on Earth; geologists have discovered more than 25 similar localities where unusual volcanism is taking place, commonly far from recognized plate boundaries (Figure 9.29). Hotspots clearly represent a major geological feature where a considerable amount of heat is rising from Earth's interior.

A full theory of how Earth works would seem to require a consolidation of plate tectonics and hotspot activity. One such model is a variation on the simple model of convection currents. In this view, hotspots represent major upwellings of heated material from deep in Earth's mantle. Plate movement may be largely driven by the cold, dense ocean crust sinking at subduction zones, pulling the plates along with them. In this picture, the Mid-Ocean Ridge is a relatively passive zone where the lithosphere is pulled apart and magma rises to the surface to fill in the gaps.

9.5 History of the super-continent Pangaea

Most of Earth's major surface features can be explained within the framework of plate tectonics (Figure 9.30). Great mountain ranges, volcanoes and volcanic islands, narrow oceans like the Red Sea, and rift valleys that cut through the continents are just a few of the features produced where plates collide or are pulled apart. In the past, Earth's internal forces have split super-continents apart, created new oceans, and

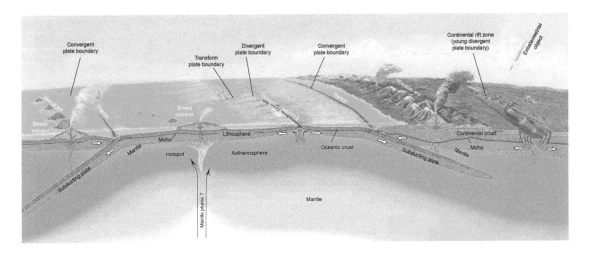

then destroyed the ocean floors by subduction. Plate tectonics occurs in a repeating cycle of opening and closing of ocean basins, accompanied by collisions of continents, that has been going on for several billion years.

Prior to the formation of Pangaea, several separate continental fragments existed. The assembly of the super-continent started some 450 million years ago, and by about 225 million years ago, Pangaea was complete. The Northern Hemisphere continents (called Laurasia) and the southern continents (called Gondwana or Gondwanaland) were joined on the west, but partly separated on the east by a large inlet called the Tethys Sea (Figure 9.31(a)).

Pangaea began to break up around 185 million years ago, with the opening of the present-day North Atlantic (Figure 9.31(b)). By 135 million years ago, we can see the beginnings of the split between Africa and South America that eventually became the South Atlantic Ocean (Figure 9.31(c)). We can also see how India split away from the other Gondwana continents about 150 million years ago to become an island continent.

By about 65 million years ago, at the time of the extinction of the dinosaurs, the Atlantic and Indian Oceans continued to widen, and India approached Asia (Figure 9.31(d)). The continued opening of the Atlantic, the collision between India and Asia, and northward motion of Australia led to the present configuration of the continents (Figure 9.31(e)).

Plate tectonics explains the major features of Earth, and the history of Earth's crust over billions of years. As plates separate, move, and collide they also affect the oceans and

FIGURE 9.30 Major features of Earth's surface explained by plate tectonics. The cross-section shows various kinds of plate boundaries and the geological features produced at those locations.

USGS

FIGURE 9.31 The breakup of the super-continent Pangaea. (a) 225 million years ago, all of the continents were a single super-continent called Pangaea. (b) Pangaea split into Laurasia and Gondwana, as the central Atlantic Ocean opened up. (c) About 135 million years ago, the South Atlantic and Indian Oceans began to open. (d) By 65 million years ago, the split between Europe and Greenland was initiated. India continued to move towards a collision with Asia. (e) Over the last 65 million years, Australia and Antarctica split apart, India collided with Asia, and the continents moved to their present positions.

USGS

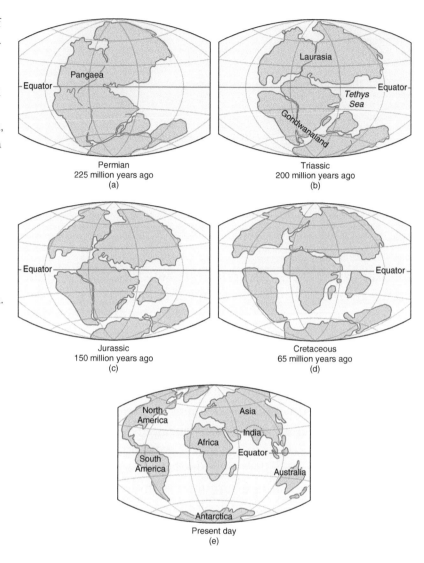

atmosphere. Another great discovery of the past few decades is that plate movement is a major cause of climate change over timescales of millions of years, and that these climate changes have greatly influenced the history of life on Earth.

9.6 Summary

By the end of this chapter you should be familiar with the following concepts and topics:

The changing face of an active habitable planet
From Pangaea to the continents we see today: the theory of
plate tectonics
The birth of a theory: match between the coastlines of
South America and Africa – Alfred Wegener
The distribution of fossil plants and animals across continents
The distribution of rock types and ages across continents
The record of changes in Earth's magnetic field
The ocean floor: Mid-Ocean Ridge, magnetic stripes, age
of rocks
The lithospheric plate boundaries
Diverging plates and opening of a new sea
Converging plates; ocean–ocean; ocean–continent;
continent–continent sliding plates Hotspots

Questions
9.1. What are the differences between ocean crust and
continental crust?
9.2. What is the lithosphere? Why is the lithosphere brittle
and not plastic like the layer underlying it?
9.3. What are the major pieces of evidence for continental
drift?
9.4. What is the origin of the Mid-Ocean Ridge?
9.5. Where are the youngest and oldest rocks in the ocean
floor? How old is the oldest ocean floor?
9.6. How are the magnetic stripes on the ocean floor
produced?
9.7. What are the major types of plate boundaries?
9.8. What are "hotspots," and how is their origin related to
plate tectonics?
9.9. What was Pangaea? When did it break apart?

CHAPTER 10

Climate change and the evolution of life

Climate changes and Earth history

Most life as we know it can exist only within a narrow range of temperatures, and changing climate has been a major driving force in the evolution of life on Earth. Historical and geological records show that the global temperature on Earth has fluctuated widely throughout history. On scales of hundreds of millions of years, the planet oscillated between warm intervals and major ice ages when the planet may have been entirely covered by ice. Within each major ice age, the climate record shows a regular advance and retreat of the ice sheets. In the shorter term our climate has been marked by relatively small changes in global temperature. Throughout history, climatic changes have determined the destiny of life, allowing certain life forms to flourish whilst others are pushed to extinction.

In this chapter we will look at the three major scales on which Earth's climate shows fluctuations. The occurrence of major ice ages is known to be connected to plate movements, which cause changes to the surface of Earth, and we will see how tectonics are directly involved in heating or cooling a planet. We will then move on to look at the causes of climate changes within each major ice age, which cause a regular advance and retreat of the ice sheets, and how this pattern is linked to Earth's orbit around the Sun. On a much shorter term, climate change has had significant effects on human activities, and direct measurement of temperatures has shown distinct trends. We will look at the causes of short-term climate change, both natural and human-related, and what might be in store for us in the future.

10.1 Timescales of climate change

Long-term climatic change shows fluctuations over tens to hundreds of millions of years. These long-term changes involve swings in global temperature of as much as 10 °C (about 20 °F). At the low points of these long-term cycles, ice sheets cover large areas of the planet. These times are known as *major ice ages*. At the high points of these fluctuations, Earth experiences long intervals when the entire planet is considerably warmer than at present, and large ice sheets are absent from the globe.

Within a major ice age, temperature fluctuations of roughly 6 °C (about 10 °F) occur in cycles of tens to hundreds of thousands of years. These variations produce *glacial* and *interglacial* periods, i.e., times within an ice age when ice sheets alternately advance and retreat across the continents.

Within a glacial or interglacial period, smaller changes of about 0.5 °C (1 °F) occur with timescales of about a hundred to a thousand years. Several such episodes of climatic change have been recorded in historic times. A shift of as little as 0.5 °C in the average temperature of the globe can have a significant impact on agriculture, transportation, and other human activities.

Previous page: Canada glacier in the Taylor Valley, Victoria Land, Antarctica.
    Tracey Stela, National Science Foundation.

10.2 Climate change over hundreds of millions of years

Continental ice sheets leave behind telltale evidence in the form of scratches and grooves, gouged out by the moving ice, and characteristic deposits containing large boulders moved by the ice. These kinds of evidence in the geological record clearly indicate that major ice ages occurred from about 300 to 260 million years ago (the Late Paleozoic Ice Age) and about 440 million years ago (the Ordovician Ice Age). There is also evidence for the occurrence of a number of extensive earlier ice ages (for example, the Sturtian Glaciation at 710 million years ago and the Marinoan Glaciation at about 640 million years ago). There are also traces of an earlier glaciation roughly 2.3 billion years ago (Figure 10.1). During these Precambrian glaciations, the planet may have been entirely or almost entirely covered by ice. These times have been named Snowball Earth episodes, and are the subject of intense study.

10.2.1 Plate tectonics and continental drift: A cause of major ice ages

For a long time, geologists puzzled over the causes of the long-term oscillations of the global climate between warm intervals and major ice ages. With the acceptance of plate tectonic theory, it became clear that there was a connection between major ice ages and plate movements. The explanation involves a chain of circumstances. During times of very active ocean-floor spreading, new ocean crust is generated at a rapid rate. The new crust is hot and buoyant, causing the Mid-Ocean Ridges to become elevated.[1] This uplift of the ocean floor displaces water from the ocean basins on to the continents, forming shallow

---

[1] As the ocean lithosphere cools by conductive heat loss, it contracts and its density increases. The increase in density causes the oceanic plate to subside, and hence the depth of the ocean floor increases with its age and distance from the ridge axis. The rate of cooling, rapid at first and then slower, is proportional to the square root of the age of the sea floor. The depth of the ocean floor in relation to its age is given by:

$$\text{Depth} = 2500 \text{ meters} + 350 \sqrt{\text{age}} \text{ (millions of years)},$$

where 2500 meters (7500 feet) is the average ocean depth at the crest of the Mid-Ocean Ridge.

(a)

(b)

continental seas. Thus, the amount of exposed land area decreases, and the amount of water-covered area increases.

How does the amount of exposed land affect the climate? Land reflects more of the incoming solar radiation back to space than ocean water. Land reflects from 15% for vegetation-covered areas to 35% for bare desert regions. By contrast, ocean water reflects only about 7% of the incoming sunlight back to space, i.e., it absorbs much more of the solar energy than the land does. Thus, the larger the area of Earth's surface covered by water, the more solar energy is absorbed. This tends to warm the planet.

When sea-floor spreading is slow, less of the new warm ocean floor is formed, the ocean floor cools and subsides, and water drains from the continental seas back into the deepening ocean basins. As a result, the amount of dry land increases. This means that more solar energy is reflected, less is absorbed, and the planet is cooled. These conditions are favorable for the onset of a major ice age.

The movement of the continents is another factor in the initiation of major ice ages. When continents drift to polar latitudes, snow and ice can begin to accumulate. The snow and ice reflect an especially large percentage of incoming solar radiation back to space (65% for snow, up to 80% for ice), further cooling Earth, and providing an impetus to further growth of glaciers.

During the most recent or Pleistocene Ice Age (the last 2 million years), there is a concentration of land area in the northern latitudes where the ice sheets wax and wane. During

FIGURE 10.1 Evidence of past major ice ages.
(a) Glacially scratched and scoured bedrock surface about 430 million years old. This outcrop is located today in the Sahara Desert in Algeria, but 430 million years ago this region was near the South Pole.
   R. W. Fairbridge
(b) Glacial deposits from southern Canada dating from about 2.3 billion years ago.
   M. R. Rampino

FIGURE 10.2 The carbonate–silicate geochemical cycle. Chemical reactions that weather rock minerals remove carbon dioxide from the atmosphere. This carbon dioxide (CO_2) is eventually locked away as carbonates in ocean sediments. Subduction and heating of these sediments release volcanic carbon dioxide back into the atmosphere, as does volcanism at the Mid-Ocean Ridge.

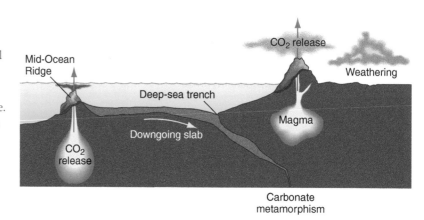

the Late Paleozoic Ice Age of about 300 million years ago, a large part of the super-continent Pangaea was concentrated near the South Pole, and the Southern Hemisphere land masses were covered by continental glaciers.

The concentration of land at high (polar) latitudes favors the formation and growth of great ice sheets, as noted, reflecting more incoming solar energy, and further cooling the planet. This generates a *positive feedback* effect: formation of ice sheets further cools the planet because ice reflects 80% of solar energy. Cooling of the planet leads to the formation of more ice. Formation of additional ice cools the planet even more, leading to formation of more ice, and so on.

A third important control on long-term climate is the amount of carbon dioxide (CO_2) in the atmosphere. Just as in the Venus atmosphere, carbon dioxide is a greenhouse gas in Earth's atmosphere, though it is only present in traces (about 4 parts of carbon dioxide for every 10 000 parts of air). Through a chain of circumstances, plate tectonics also influences the amount of carbon dioxide in Earth's atmosphere (Figure 10.2). In the first step, carbon dioxide in the atmosphere reacts with surface rocks causing weathering of silicate minerals (through the same chemical reactions discussed in the chapter on Venus). This converts the carbon dioxide to carbonate, e.g., calcium carbonate, $CaCO_3$, and removes it from the atmosphere. (The carbonate is carried in a dissolved form into the ocean, where it is eventually precipitated out as deposits of calcium carbonate (e.g., chalk and limestone), largely in the form of the shells of marine organisms.)

However, the carbon dioxide may be returned to the atmosphere through the processes of plate tectonics. When ocean crust is subducted, the calcium carbonate-rich sediments on the ocean floor are heated, and carbon dioxide is released. Some of this carbon dioxide returns to the atmosphere through subduction-zone volcanoes. The rate at which this recycling of carbon dioxide occurs depends upon several factors. The faster the rate of subduction, which varies with the rate of sea-floor spreading, the faster the recycling of carbon dioxide. The area of the continents exposed to weathering also affects the recycling rate of carbon dioxide, so the greater the area of dry land, the faster the removal of carbon dioxide from the atmosphere by weathering; and the less dry land, the slower the removal of carbon dioxide. This is another reason why low sea level leads to cool temperatures.

The best current estimates of the carbon dioxide content of the atmosphere over the last 600 million years, based on the balance between the supply and removal of atmospheric carbon dioxide are shown in Figure 10.3. These estimates range from today's values to almost 20 times the present carbon dioxide content of the atmosphere.

The amount of carbon dioxide in the atmosphere depends primarily on the rates of plate motion. When sea-floor spreading and hence subduction are fast, volcanism at the Mid-Ocean Ridge releases carbon dioxide, and the chalky

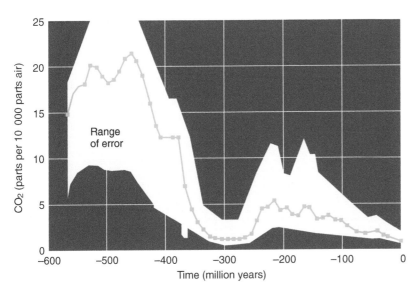

FIGURE 10.3 Carbon dioxide (CO_2) content of the atmosphere for the last 600 million years as estimated from a computer model using geological information on the supply of carbon dioxide to the atmosphere from volcanism and its removal by deposition of calcium carbonate. The range of error is based on geological evidence.

sediments formed in the oceans are rapidly dragged down and heated in the subduction zones, also releasing carbon dioxide to the atmosphere at a greater rate. Since sea level is also high at these times, there is less exposed land, and rock weathering is less efficient in removing atmospheric carbon dioxide, hence the climate remains warm.

Thus, rapid sea-floor spreading creates conditions for a warm climate by enhancing the amount of atmospheric carbon dioxide and the greenhouse effect. This is in addition to the effect of rapid sea-floor spreading in warming the planet by raising sea levels.

When sea-floor spreading is slow, the amount of atmospheric carbon dioxide decreases because subduction of ocean-floor carbonates is slowed. With less carbon dioxide returned to the atmosphere by volcanism, the greenhouse effect is smaller, and the planet is cooled. This cooling effect is in addition to the cooling caused by a drop in sea level, as described above. As global temperature decreases, ice sheets begin to grow, marking the onset of a major ice age.

10.3 Snowball Earth

The most ancient glacial deposits, dating from about 600 million years ago back to more than 2 billion years ago, are widespread, and some are found in regions that were near the Equator at those times in the past. These findings of ice sheets in the tropics suggest that the entire planet, or at least a major portion of it, must have been frozen over. These times of worldwide glaciations or *Snowball Earth* episodes (Figure 10.4) may have been initiated when extensive weathering of continental rocks removed an inordinate amount of carbon dioxide from the atmosphere. As the carbon dioxide levels fell and the climate grew cooler, ice sheets formed and advanced toward lower latitudes, cooling the global climate further by their high reflectivity. The end result was periods of almost total refrigeration of the planet.

The slow buildup of atmospheric carbon dioxide from volcanic eruptions would eventually have warmed the planet sufficiently to melt the global ice cover. After another long

FIGURE 10.4 Deposits from a Snowball Earth episode more than 600 million years ago in Namibia (southwest Africa). This region was near the Equator at that time, suggesting that ice may have covered the entire globe.
Paul Hoffman and Daniel P. Schrag, Harvard University

period during which carbon dioxide was again removed from the atmosphere by rock weathering, Earth would have grown cold enough for ice sheets to reform. These ice sheets again spread rapidly to cover a large portion of the planet. In this way, cycles of global glaciation and deglaciation alternated during each Snowball Earth episode. Many scientists believe that these drastic changes in climate could have created periods of crisis for early life forms. The Snowball Earth episodes ended abruptly about 600 million years ago.

Why have they not recurred? The answer may be that the episodes of planet-wide glaciation were possible early in Earth's history because the early Sun was dimmer and gave off less energy. Astronomers have determined that as the Sun ages it grows brighter. This is because as helium builds up in the Sun's core, the core density increases, and the temperature rises. A hotter core means faster conversion of protons into helium, and thus more energy is released.

10.4 Episodes of climate change lasting tens to hundreds of thousands of years

In the long term, we are now in the midst of a major ice age. Ice sheets formed in Antarctica as early as about 35 million years ago, and during the last 2.5 million years (called the Pleistocene Ice Age) the cooling became so severe that glaciers periodically advanced and retreated across North America and Eurasia. As recently as 18 000 years ago, ice sheets reached their maximum extent and covered most of northern Europe and north America (Figure 10.5).

The temperature record of the last 2.5 million years is shown in Figure 10.6, in which we can see the cooling trend that culminated in the most severe glaciations during the last 500 000 years, accompanied by extreme shifts from cold glacial intervals to warm interglacial epochs.

The climate record of the last 500 000 years shows the progression of cold glacial periods during which ice sheets expand, alternating with warmer interglacial periods in which the ice sheets retreat (Figure 10.7). The succession of glacials and interglacials represents changes of only about 6 °C (about 10 °F) in average global temperature. Note that the warm peaks – times when global temperatures were as high as they are today – make up only a small fraction of the last few hundred thousand years. Note also that none of the interglacial periods of relatively warm, ice-free climate has lasted more than about 10 000 years. The last glacial period ended about 11 000 years ago, so the next glacial epoch would seem to be about 1000 years overdue.

What is the cause of the regular advance and retreat of the ice sheets every 10 000 to 100 000 years within a major ice age? Today, the astronomical explanation put forth by the Serbian mathematician Milutin Milankovitch in the 1920s is generally accepted. Milankovitch suggested that the glacial and interglacial epochs were caused by cyclic variations in Earth's orbit around the Sun, and in the tilt of Earth's axis. These variations affect the amount of solar energy falling on the land areas of the Northern Hemisphere, and especially the distribution of that energy over the seasons.

The perturbations of Earth's orbit are caused primarily by the gravitational pull of the planet Jupiter and Earth's Moon on Earth. Earth's orbit varies in the following ways, as a consequence of these perturbing forces.

The *shape* of Earth's orbit around the Sun changes from circular to elliptical or "egg-shaped" and back every 105 000

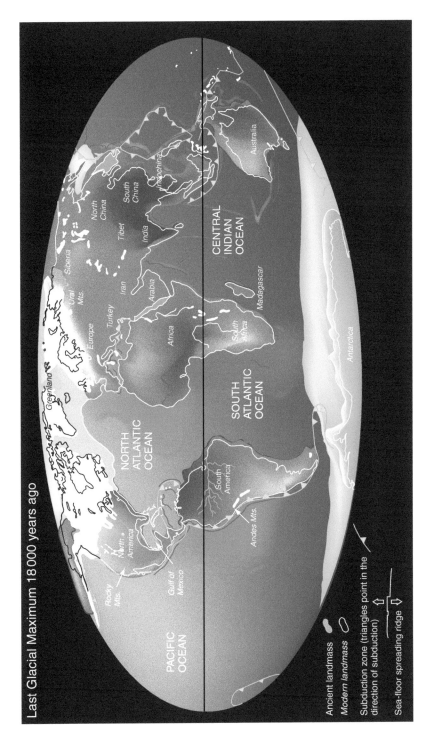

FIGURE 10.5 The last glacial period culminated in ice sheets that covered Northern Europe and North America down to the latitude of New York City about 18 000 years ago.
Christopher R. Scotese

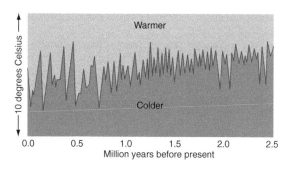

FIGURE 10.6 Fluctuations in global temperatures over the last 2.5 million years. A general cooling trend can be seen, as well as oscillations representing cold glacial and warmer interglacial intervals. Note the increase in the amplitude of the glacial/interglacial swings about 700 000 years ago.

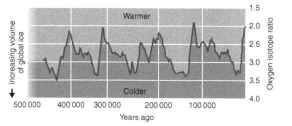

FIGURE 10.7 Glacial and interglacial climate over the last 500 000 years. Glacial periods typically last almost 100 000 years, whereas interglacials last only about 10 000 years.

years (Figure 10.8(a)). In the most extreme case, however, the egg-shape is only about 10% from circular. At the present time, the orbit is about 2% from circular, and is becoming even more circular.

Earth's axis of rotation precesses every 23 000 years, i.e., Earth wobbles like a top, taking 23 000 years to complete one wobble (Figure 10.8(b)). This effect is mainly caused by the Moon's gravitational pull. The precession of Earth's axis affects the direction in space in which Earth is tilted at any time of the year. Today Earth's axis points almost exactly to the star Polaris. Five thousand years ago it pointed to the star Thuban, and 11 500 years from now the axis will be pointing at Vega. The effect of precession is to reverse the hemisphere experiencing summer at the time of closest approach of Earth to the Sun. Today, Earth is closest to the Sun in January, but 11 500 years ago Earth was closest to the Sun in July, corresponding to Northern Hemisphere summer.

The angle of *tilt* of Earth's axis also changes, from about 22.5° to 24.5° and back again every 41 000 years (Figure 10.8 (c)). The present angle of tilt of the axis is 23.5°, and this angle is currently decreasing. The smaller the angle, the smaller the temperature difference between the seasons. Hence, winters are milder and summers cooler.

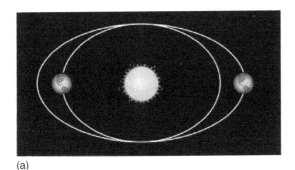
(a)

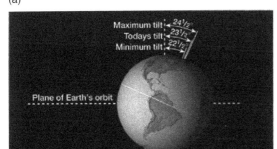
(b)

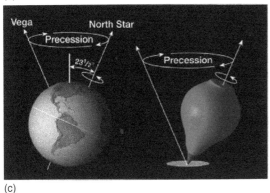
(c)

FIGURE 10.8 Changes in Earth's orbit.
(a) Earth's orbit changes shape from egg-shaped (ellipse) to circular every 105 000 years (not to scale).
(b)Tilt changes from 22.5° to 24.5° every 41 000 years.
(c) Precession of Earth's axis of rotation every 23 000 years, in which Earth acts like a wobbling top.

Terbuck, E. J. & Lutgens, F. K., *Earth Science*, 9th, © 2000. Reproduced by permission of Pearson Education, Inc., Upper Saddle River, New Jersey

These circumstances change the intensity of solar radiation striking the land areas in the Northern Hemisphere during the various seasons, and therefore change the rate of snow and ice accumulation on the northern continents. To see why this is so, consider the present situation on Earth. At the present time, because of the somewhat elliptical shape of Earth's orbit, Earth's distance from the Sun varies by 3 million miles during the course of a year. In what part of the year is Earth closest to the Sun? At present this occurs in the month of January, when the Northern Hemisphere is tilted away from the Sun (Figure 10.9).

FIGURE 10.9 Conditions for mild winters and cool summers in the Northern Hemisphere (reduced seasonal variation). (Left) Earth's axis is tilted *away* from the Sun in the Northern Hemisphere when Earth is closest to the Sun in its orbit. (Right) axis is tilted toward the Sun when Earth is farthest from the Sun in its orbit. Mild seasonality and glaciation begins. This is the situation now. We are headed into a glacial period. (Not to scale. The actual difference in winter/summer distance from the Sun is about 3%.)

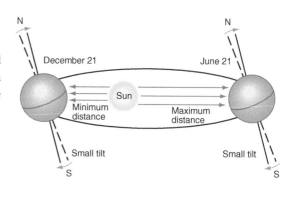

FIGURE 10.10 Conditions for cold winters and hot summers in the Northern Hemisphere (extreme seasonal variation). (Left) Earth's axis is tilted *toward* the Sun in the Northern Hemisphere when Earth is closest to Sun in its orbit. (Right) axis is tilted *away* from the Sun in the Northern Hemisphere when Earth is farthest from the Sun. During times of extreme seasonality, ice sheets melt, leading to an interglacial period. (Not to scale. The actual difference in winter/summer distance from the Sun is about 3%.)

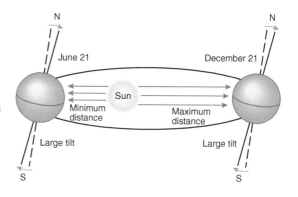

Because the Northern Hemisphere is tilted away from the Sun, the Sun's rays are incident on the Northern Hemisphere at an oblique angle, and we receive less solar heat. That is why the winter season prevails in the Northern Hemisphere (remember, in Australia, January is mid-summer). However, because the Earth is relatively close to the Sun at that time, the Northern Hemisphere winters tend to be relatively mild. At the same time, the summers in the Northern Hemisphere tend to be relatively cool.

Now consider the situation roughly 11 500 years in the future (one-half the precession cycle), when, because of the precession of Earth's axis, the tilt of the axis is in the opposite direction in space. Now the Northern Hemisphere is tilted *toward* the Sun, and therefore experiencing summer, at just the time when Earth is closest to the Sun (Figure 10.10). Therefore, Northern Hemisphere summers should be relatively hot. Six months later, on the other side of Earth's orbit, the Northern Hemisphere is tilted away from the Sun (winter) just when Earth is farthest from the Sun. Therefore, the winters should be quite cold.

Present conditions on Earth are such as to make for a succession of mild winters and cool summers. Eleven thousand years from now, conditions will be such as to produce a succession of cold winters and hot summers in the Northern Hemisphere.

The effects we have just described depend on how egg-shaped (elliptical) the orbit is. These effects are greatest when the orbit is very egg-shaped and the Sun–Earth distance is accordingly very different in January than in June. The effects also depend on the tilt of the axis. When the tilt of the axis is at its greatest value of 24.5°, the effects are enhanced, but they are diminished when the tilt is less. (If the tilt were zero, i.e., Earth's axis were always exactly perpendicular to the plane of its orbit, there would be essentially no seasons on our planet.)

How do these variations in Earth's orbit lead to the alternate growth and retreat of ice sheets? When the orbit is such that Northern Hemisphere winters are extremely cold, as in Figure 10.10, it would seem that snow and ice accumulation should be most rapid, i.e., a glacial period should result. However, the data show the opposite: The ice starts to accumulate during the periods in which the Northern Hemisphere winters are relatively mild and the summers are cool as in Figure 10.9.

Heavy snow will occur in most winters, but cool summers mean that the winter snow that falls is more likely to last through the summer, accumulating from year to year to become permanent snowfields. The snowfields are converted to ice sheets by the pressure of the accumulating snow, and eventually the thick ice flows outward to form continental glaciers.

The growth of the ice sheets actually speeds up the cooling of the planet. Ice, being white, reflects most of the incident solar energy back to space, whereas water and land absorb most of this energy. The presence of snow and ice cover, therefore, means that Earth is absorbing less energy from the Sun. This effect leads to further growth of the ice sheets;

when the ice sheets become larger, Earth absorbs still less solar heat; and so on. For this reason, the ice sheets increase in size very rapidly once they start to grow, until they cover a considerable portion of the land area of Earth.

Another process that amplifies the global cooling involves atmospheric carbon dioxide and methane. We know from analyses of air bubbles trapped inside ancient layers of polar ice that during glacial periods the atmospheric carbon dioxide and methane content were lower than during interglacials. This reduction in greenhouse gases would cause a further cooling of the climate during the glacial periods.

Is there any evidence that this explanation of the ice ages is correct? In the mid 1970s, a group of researchers found that the record of glacial and interglacial periods in the last 500 000 years (as shown in Figure 10.8) could be broken down into cycles with clear periods of 23 000, 41 000, and 100 000 years. These periods agree so well with the cycles of changes in Earth's orbit listed above as to leave little doubt that the two phenomena – the orbital changes and the glacial/interglacial cycles – are related.

10.5 Short-term climate change: The last 10 000 years

We emerged from the last glacial period about 11 000 years ago. Our climate since that time has been marked by relatively small changes of 0.5 to 2 °C (about 1 to 4 °F) in global temperature (Figure 10.11). However, these seemingly small changes in climate have had significant effects on human activities.

The warming in the present interglacial reached a peak about 6000 years ago, in about 4000 BC. This period of mild climate may have led to large agricultural yields and food surpluses. It has been suggested that the birth of civilization in Mesopotamia and Egypt was made possible by the new surpluses of food, which allowed new members of society such as administrators, teachers, and lawyers to exist in appreciable numbers for the first time.

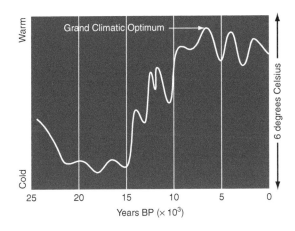

FIGURE 10.11 The last 25 000 years of climate, including the present interglacial, showing broad climate trends, notably the Grand Climatic Optimum of about 6000 years ago. That was the warmest time within the present interglacial. Twentieth century climate is shown in Figure 10.12.

This so-called *Grand Climatic Optimum* of about 4000 BC. was followed by a general cooling trend. Another period of pronounced climate change occurred between AD 800 and AD 1300, when Earth experienced a warming that has been called the *Medieval Warm Period*. During that period, Northern Hemisphere temperatures increased by a few tenths of a degree Celsius. However, that small change had far-reaching effects, for the period of moderately warm weather expanded the growing season and the amount of arable land in the Scandinavian countries. The population in these countries expanded accordingly, and the Vikings began to push east, south, and west. They captured Paris in AD 845, and reached the area of Moscow in the same period.

10.5.1 The Medieval Warm Period

In their drive to the west, the Vikings sailed across the North Atlantic and settled in Iceland, Greenland, and North America (Newfoundland, named Vinland by the Norse because they found grapes growing there) around AD 1000. The Vikings were able to accomplish this because the North Atlantic was relatively calm and free of sea ice during the Medieval Warm Period. The Norse also set up colonies in coastal Greenland and raised crops in an area that today is a frozen ice-covered waste. But in the fourteenth century, as the climate worsened, the North Atlantic colonies were cut off from their homeland by sea ice and increasing storminess. As the cold weather closed in, the Greenland colony became extinct by about AD 1400, leaving America to be rediscovered by Columbus.

10.5.2 The Little Ice Age

A cooling trend is recorded from the fifteenth to the late nineteenth centuries as the world sank into the *Little Ice Age*. Average temperatures in the Northern Hemisphere dropped by only a few tenths of a degree Celsius, but average winter temperatures decreased as much as 20 °C (about 40 °F). These small changes led to a shortened growing season and contributed to severe famines and social unrest in Europe. In the late 1500s, for example, several million people in France and neighboring countries starved to death in climate-related famines.

Historical records reveal that the winters of the time were extremely severe. For example, snowfall in Europe was quite heavy, spring thaws were late, and major rivers were frozen over all winter. Popular frost fairs were held each year on the thick ice cover of the River Thames in England. In recent times, the Thames remains ice-free all winter long.

Temperatures began to rise in the 1700s, dropped somewhat around the beginning of the nineteenth century, and stayed rather cool for much of the 1800s. Figure 10.12 shows the global climate of the last 120 years. This record represents a compilation of data from many weather stations around the globe.

Large year-to-year variations in temperature of up to 0.5 °C (about 1 °F) are evident on the graph, and these year-to-year

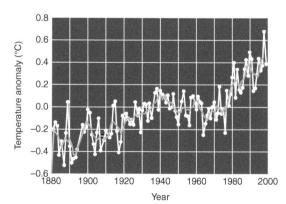

FIGURE 10.12 Average yearly global temperatures from AD 1880 to 2003 compiled by climatologists at NASA's Goddard Institute for Space Studies. The white line represents the year-to-year variations, and the blue line is the 5-year mean that shows the longer-term trends.
NASA

(a)

(b)

variations can obscure trends in climate. To overcome this problem of a noisy year-to-year record, we can look at five-year averages (shown by the solid line in Figure 10.12) over the entire period of observation. The average reveals several distinct climate trends: A general warming from the late 1800s to the 1940s, a slight cooling trend from the 1940s into the 1970s, and a warming trend from the late 1970s to the present. The long-term climate trend for the entire twentieth century was a net warming of about 0.8 °C (about 1.5 °F).

The increase in global temperature in the last 100 years has been accompanied by indications of climatic warming such as an earlier spring thaw, increase in the number of very hot summers, and the retreat of many of the world's mountain glaciers (Figure 10.13).

FIGURE 10.13 The Argentière glacier in the French Alps. (a) An etching made about 1850 showing the extent of the glacier during the Little Ice Age. (b) The same view photographed in 1966. The glacier retreated drastically up the valley in the twentieth century.

Reprinted with permission from *Understanding Climate Change: A Program for Action*, National Academy of Sciences, Washington D. C. © 1975

10.6 Causes of short-term climate change

10.6.1 Volcanic eruptions and climate

What are the causes of changes in climate that last from a few decades to a few hundred years? Several factors are believed to be important. Volcanic eruptions emit copious amounts of the gas sulfur dioxide (SO_2), which is converted to droplets of sulfuric acid (H_2SO_4) in the upper atmosphere (Figure 10.14). The sulfuric acid droplets (or *aerosols*) create a global haze that shields Earth's surface from the Sun, and cools the planet.

The global cooling caused by large explosive eruptions can be substantial, of the order of several tenths of a degree

FIGURE 10.14 The explosive eruption of Mount Pinatubo in the Philippines in 1991. This sulfur-rich volcanic eruption created a global veil of sulfuric acid aerosols in the upper atmosphere.

NOAA National Geophysical Data Center

Celsius. Several volcanic eruptions in historic times have been accompanied by noticeable decreases in temperature – for example, Tambora in Indonesia, in 1815, was followed by unusually cool weather, crop failure, and widespread famine in Europe and North America in 1816, the so-called "year without a summer." The Indonesian eruptions of Krakatau in 1883 and Mount Agung in 1963 also led to a few years of global cooling. Most recently, the eruption of Mount Pinatubo in the Philippines in 1991 produced the densest aerosol haze in 100 years, which enveloped the globe and cooled global climate by a few tenths of a degree Celsius in 1992 and 1993. However, the particles that cause the cooling disappear from the atmosphere in a few years, so that the climatic effects of even the largest eruptions are relatively brief.

10.6.2 Climate and the Sun's brightness

Changes in the Sun's brightness are another possible contributor to short-term climate change. One indication of solar activity is the presence of dark spots on the Sun's surface. *Sunspots*, first observed in modern times by Galileo in 1609, are regions of the Sun that are a few thousand degrees cooler than the gas surrounding them (about 4000 °C, compared to 6000 °C for the surface of the Sun as a whole) so they look dark compared to the areas around them. The spots commonly appear on the Sun in pairs, aligned in an approximately east–west direction.

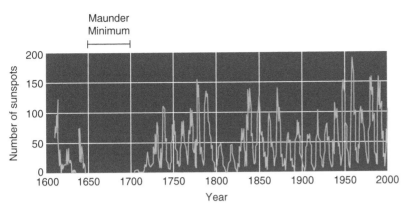

FIGURE 10.15 Numbers of sunspots AD 1600 to 2000 showing an 11-year cycle, and longer envelope of sunspot maxima. The period from about AD 1650 to 1710 – called the Maunder Minimum – was marked by a complete absence of observable sunspot activity, which correlates with the most severe part of the Little Ice Age.

Sunspots are caused by disturbances in the magnetic field of the Sun, which are caused in turn by the Sun's rapid rotation (the Sun rotates once every 27 days). They are accompanied by other solar disturbances such as *solar flares* – violent outbursts of energy and particles – and *solar prominences* – huge loops of luminous gases that erupt from the Sun and rise hundreds of thousands of miles above the solar surface.

Sunspots show an 11-year cycle of increases and decreases in the numbers of spots (Figure 10.15). Measurements of solar radiation using sensitive instruments on satellites have confirmed changes of about 0.1% in the Sun's output over the sunspot cycle. Scientists have discovered 11-year cycles in climate records such as growth rings in trees, which suggests that even small fluctuations in solar activity can affect climate. But the clearest connection between sunspots and climate has turned out to be somewhat different than originally expected. Note that the numbers of sunspots per cycle shown in Figure 10.15 are variable, changing roughly every hundred years in a fairly regular way. Furthermore, the length of a sunspot cycle, which averages 11 years, actually varies between about 9 and 15 years. Calculations by climatologists show that a small change in solar energy output could have led to the changes in climate seen in Europe during the Little Ice Age.

John Eddy at the US National Center for Atmospheric Research assembled evidence indicating that from about AD 1650 to 1710 there was almost no sunspot activity. This *Maunder Minimum* (named after an early twentieth-century astronomer who argued for its existence) coincides with the depths of the Little Ice Age. The cool temperatures in the early nineteenth century also correspond to a very low number of spots in that period.

According to Eddy, the coincidence between the Maunder Minimum and the Little Ice Age provides a clue to the cause of such century-long climate changes. It is *not* that sunspots themselves cause the climate change, but rather that the Sun's energy output is changing on a 100-year timescale, and the variations in solar energy and sunspots are both manifestations of some fundamental process going on deep within the Sun.

10.6.3 Carbon dioxide and climate: The human factor

Earth's greenhouse effect, primarily a result of carbon dioxide and water vapor in the atmosphere, keeps the surface temperature of the planet about 15 °C (27 °F) warmer than it would be without these gases. The amount of carbon dioxide and other greenhouse gases in the atmosphere has been rising over the past few centuries as a result of human activities.

The amount of carbon dioxide in the atmosphere has increased by about 40% over pre-industrial levels (Figure 10.16). Added to the increased carbon dioxide are increases in other greenhouse gases, such as methane (CH_4) from agricultural practices, nitrous oxide from nitrogen-based fertilizers, and chlorofluorocarbons (CFCs) from refrigeration and other industrial uses. Many scientists believe that a major part of the twentieth century warming has been caused by this increase in atmospheric carbon dioxide and other greenhouse gases.

What does the future hold in store? If we continue to burn coal and oil at present and projected rates, an amount of greenhouse gases equivalent to a 100% increase in carbon dioxide

FIGURE 10.16 Observed increase in atmospheric carbon dioxide from 1957, when these measurements began, to 2005. The pre-industrial atmospheric carbon dioxide, as determined from air bubbles trapped in polar ice, was only 285 parts per million. Note the increasing trend, and the seasonal cycle that is controlled by the growth and decay of Northern Hemisphere vegetation each year.
    NASA

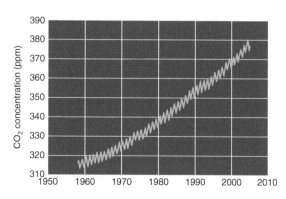

over pre-industrial values may exist in the atmosphere by the middle of the twenty-first century. Calculations suggest that this increase in carbon dioxide could cause a warming of the planet that would put stress on natural ecosystems and agriculture, increase the frequency and intensity of tropical storms, and lead to many other climatic and ecological changes.

Recently, representatives of the various nations have been meeting to try to reach agreements on limiting CO_2 emissions from fossil fuel burning in an attempt to slow global warming. Increased energy efficiency, alternative energy sources, and possible capture and sequestering of carbon dioxide emissions, along with reductions in release of other greenhouse gases would slow or reduce future global warming.

The amount of carbon dioxide in Earth's atmosphere has varied over geological time. The changing greenhouse effect and other factors have caused pressures on the forms of life, and have provided a major driving force for the evolution of life on Earth.

10.7 Summary

By the end of this chapter you should be familiar with the following concepts and topics:

Fluctuations of Earth's climate
Timescales of climate change
The occurrence of major ice ages
   Their connection to plate movements
   Changes to the surface of Earth
The causes of climate changes within each major ice age
   Cause of a regular advance and retreat of the ice sheets
   Linkage with Earth's orbit around the Sun
Short-term climate change and its effects on human activities
   Natural causes of short-term climate change: volcanic
     eruptions and the Sun's brightness
   Human-related causes of short-term climate change:
     carbon dioxide
What is in store for us in the future?

Questions

10.1. Describe three kinds of changes in Earth's orbit during the last 300 000 years.

10.2. How do these changes affect global climate?

10.3. Explain the relationship between volcanic eruptions and climate.

10.4. What is the effect of changes in the relative amounts of land and sea through time on the climate?

10.5. What are the predicted effects of fossil-fuel burning on the atmosphere and climate?

PART IV

Origin and history of life on Earth

CHAPTER 11

The origin and evolution of life

Life's beginnings and Darwin's theory of evolution

Earth began its existence 4.6 billion years ago, circling the newborn Sun. Formed out of inert atoms of gas and grains of dust, and heavily bombarded by asteroids and comets, our planet was surely just a sterile body of rock. But sometime during the first billion years of the planet's existence, life appeared. The earliest forms of life were very simple, but during the billions of years that followed they developed into the rich variety of plants and animals that now live on Earth. What guided the course of evolution on this planet from the first primitive organisms to the complicated creatures of today? Is there a law in nature that determines the forms of life? Understanding how life originated and evolved on Earth is essential if we are to contemplate the existence of life elsewhere in the Universe.

Millions of species of life exist on the planet, from simple bacteria to complex mammals. How did this enormous diversity arise? In this chapter we start from the very simplest building blocks of life, which play a central role in life's processes. We will learn how carbon compounds are built into the proteins from which all life is constructed. DNA is the most important molecule in every living organism, containing the genetic code for life. We will be introduced to its remarkable complexity, and learn how it can replicate itself to pass the code on to future generations. We will then turn to Charles Darwin, and how his observations of living organisms during his voyage around the world led him to the conclusion that species evolve.

11.1 What is life?

Consider a self-replicating machine. A machine can be programmed with a set of instructions for constructing its duplicate. When the duplicate is completed, the original machine must make a copy of the instructions, insert the copy into the new machine's computer, and turn it on. The two machines then go on making copies of themselves, so that eventually (and as long as the supply of parts continues) a population of these self-replicating machines is produced.

This kind of self-replicating machine was originally conceived by the mathematician John von Neumann. One problem with these machines, however, is that they appear to be unchanging. Left alone, the von Neumann machines should go on building exact copies of themselves forever. It seems that if we were to return after millions of years, we would find that the population of machines is still uniform. Life on Earth, by contrast, includes a remarkable range of living things as different as bacteria and human beings. Millions of species of life exist on the planet, from the simple to the complex. Thus, although we have created a machine that can reproduce, we have not produced the variety or complexity that are a critical manifestation of life on Earth. In fact, most biologists agree that the two essential qualities that differentiate life from

Previous page: Detail from the ceiling of the Sistine Chapel by Michelangelo Buonarotti.

non-life are reproduction and life's ability to change in response to changes in the environment.

As it turns out, however, the self-replicating machines have the ability to change and evolve, and they can do so in a manner similar to living organisms. The ability to evolve comes out of the process of copying the instructions to pass on to the newly built machine. While this duplication will be accurate most of the time, no process of copying is infallible. Every now and then a copy error must occur, and the instructions that are passed on will be changed in some way.

In most cases, the copy error will be so great as to prevent the new machine from building a working copy of itself, and that line of machines will become extinct.

But every so often, a random change in the instruction program will lead to a small change in the machine being constructed If the change happens by chance to be a small improvement in the workings of the machine, then that line of machine may build copies more efficiently; i.e., it will have greater reproductive success. Such machines will come to dominate the population. The variations produced by such copy errors provide the raw material for evolution.

The second factor that comes into play is natural selection. The environment can change to favor one variety of machines over another. Since we cannot predict the nature of the random copy errors or the vagaries of environmental change, we cannot predict precisely what forms the machines will eventually take. Thus, the key to evolution is replication with mistakes, producing variations in the forms of life that are either more or less successful in the local environment. This, it turns out, is the essence of life.

11.2 The basic building blocks of life

Life on Earth is based on the carbon atom, and the unique ability of carbon to bond with other carbon atoms to form long chainlike molecules. Carbon also bonds readily with atoms of other common elements such as hydrogen, oxygen, nitrogen, phosphorus, and sulfur. Double or triple bonds can also form between carbon atoms. In this way, carbon atoms

FIGURE 11.1(A) The structure of the 20 amino acids. All of the amino acids have a similar basic structure, but with different side branches (blue areas).

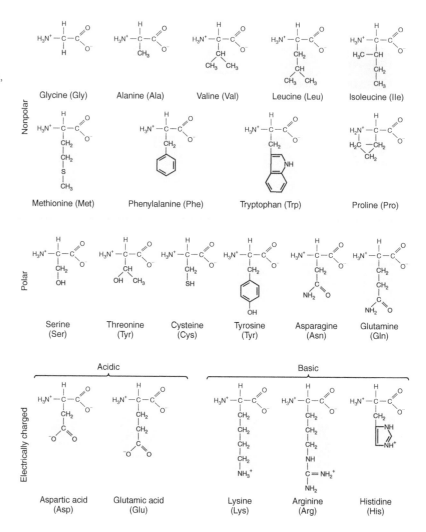

can form rings and long chains that have additional bond positions available for linking with other atoms.

Two particular kinds of carbon compounds – *amino acids* and *nucleotide bases* – play a central role in life's processes today (Figure 11.1).[1] Twenty different kinds of amino acids and five different kinds of nucleotide bases occur in living things. Each amino acid or nucleotide base is a molecule made up of approximately 10 to 30 atoms of hydrogen, nitrogen, oxygen, and carbon, held together by chemical bonds.

[1] Other molecules are also important. For example, lipids are essential in membranes, which today form the essential compartment that encloses cells. Energy release by intermediary metabolism, involving sugars and other molecules, is also essential.

Pyrimidines

Cytosine
C

Thymine
(in DNA)
T

Uracil
(in RNA)
U

Purines

Adenine
A

Guanine
G

FIGURE 11.1(B) The structure of the bases of the five nucleotides, showing the basic ring structure with single and double bonds between carbon atoms and nitrogen atoms. Cytosine, thymine, adenine, and guanine occur in DNA. Uracil is found only in RNA.

Nucleotide bases combine with phosphates and sugars to form nucleotides.

11.3 Proteins

Within the cell, the amino acids (Figure 11.1(a)) are linked together into very large chains called *proteins*. These chains of amino acids are produced because a hydroxyl ion (OH^-) at the end of one amino acid molecule can combine with the hydrogen ion (H^+) at the end of another amino acid molecule. This forms a water molecule (H_2O) (Figure 11.2(a)), which is released, leaving the two ends of the amino acids stuck together (Figure 11.2(b)). This *peptide bond*, as it is called, links the amino acids into a chain called a *polypeptide*. Polypeptides can then be extended into the much longer chains that constitute protein molecules.

Some proteins make up the structural elements of the living organism. These structural proteins are like the steel framework and walls of a building. Other proteins, called *active proteins*, carry out the organism's functions such as transportation of critical materials. Another type of protein

FIGURE 11.2 The linking of amino acids to form a polypeptide – the beginnings of a protein. The amino acids are linked together by the formation of a water molecule that is subsequently released.

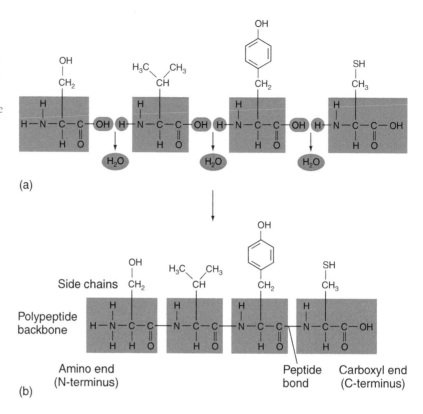

(a)

(b)

Side chains

Polypeptide backbone

Amino end (N-terminus)

Peptide bond

Carboxyl end (C-terminus)

is an *enzyme*. Many kinds of enzymes exist; each promotes and regulates one of the many chemical reactions that are necessary to sustain the life of the organism.

All proteins in all forms of life, plant and animal, are constructed out of the same basic set of 20 amino acids. One protein differs from another in the order in which its constituent amino acids are linked together (Figure 11.3). However, these differences are all-important. The distinction between proteins of different functions depends largely on the differences in the sequence of amino acids in the protein molecule.

In a typical protein, the long chain of amino acids is coiled into a helix, a shape similar to a coiled telephone cord. The helix then folds into a crumpled and cross-linked shape because some of the amino acids have places on their sides where weak bonds can form with other amino acids (Figure 11.3). The final protein molecule has a complex shape (e.g., spiral ladders called helices, and pleated sheets), defined by its particular sequences of amino acids. The complex shape of the protein molecule is the key to its function in living things. For example, *hemoglobin* (Figure 11.4) is a complex of four protein sub-units that has just the right shape and structure to hold oxygen atoms, which can then be transported from one place to another in a cell or organism.

11.4 DNA

There are five different kinds of nucleotide bases (Figure 11.1(b)). Nucleotide bases, when combined with phosphates and sugars to form nucleotides, are joined together within the cell to form very long chains, called nucleic acids. The most important type of nucleic acid is called deoxyribonucleic acid, or DNA for short. Only four of the five important nucleotide bases (adenine, guanine, cytosine, and thymine) enter into the structure of DNA. The fifth nucleotide base (uracil) belongs to another type of nucleic acid, called ribonucleic acid or RNA.

DNA is the largest molecule known, containing, in complex organisms such as humans, as many as 10 billion separate atoms. The size of the DNA molecule reflects the complexity and importance of its functions in the living cell. The DNA molecule is the most important molecule in every living organism, because the DNA molecule contains the master plan for the organism (Figure 11.5).

The structure of the DNA molecule was worked out in 1953 by the team of James D. Watson and Francis Crick working at Cambridge University. This discovery was one of the most important single scientific events of the twentieth century. The basic structure of DNA resembles a twisted ladder. The rails of the ladder are made of alternating phosphates and sugars, and the rungs of the ladder are made up of pairs of nucleotide bases. In DNA, an adenine molecule (A) on one strand is always matched with thymine (T) on the other, and a guanine (G) is always matched with a cytosine (C). Thus, the sequence of nucleotide bases on one strand precisely determines the complementary sequence of nucleotide bases on the other.

How does DNA control the assembly of proteins in the cell? The separate amino acids and nucleotide bases move freely in

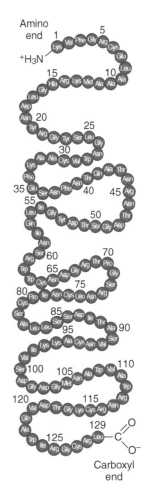

FIGURE 11.3 Protein structure. A protein is composed of a long chain of amino acids. These chains are then folded and refolded to make the final complex form of the active protein. The shape of the protein defines its role as an enzyme that can promote some reaction crucial for life.

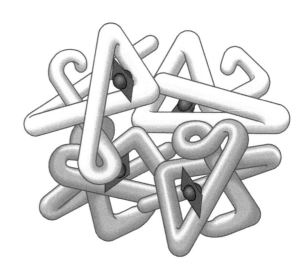

FIGURE 11.4 The hemoglobin molecule. The heme component (shown in red) of this complex of four protein sub-units (in blue) contains an iron atom (in brown) that can bind to oxygen atoms.

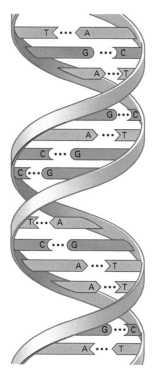

FIGURE 11.5 The structure of DNA. DNA resembles a ladder; each rung of the ladder is a linked pair of nucleotide bases, represented by the symbols: A, G, C, and T. The ladder is twisted into a double spiral, called a double helix.

the fluid of the cell. In the first step, with the help of special proteins, a portion of the DNA molecule unzips along the middle of the rungs. With the help of other special proteins, nucleotide bases are assembled along one of the segments of the unzipped DNA. These nucleotide bases produce a long strand of RNA called *messenger RNA*.

In the second step, the messenger RNA detaches itself from the master DNA chain, and is carried off into the cell; it is a messenger that carries instructions from the DNA into the body of the cell for the assembly of one particular kind of protein (Figure 11.6). In the third step, another type of RNA molecule enters the picture. This molecule is called *transfer RNA* and serves as a connecting link, bringing the amino acids in the fluid of the cell to the appropriate places alongside the messenger RNA molecule. Each kind of transfer RNA attracts one and only one of the 20 amino acids. When the right kind of amino acid comes into contact with the end of the transfer RNA molecule designed for that particular amino acid, it is held fast there.[2]

At the other end of the transfer RNA molecule is a set of nucleotide bases that can attach only to their complementary nucleotide base bases along the length of the messenger RNA. When the transfer RNA takes its designated place along the

[2] These processes take place on a structure in the cell called the ribosome. The ribosome may be considered as the factory where proteins are manufactured. The ribosome is made of about 50 protein and RNA subunits, and it binds and holds messenger and transfer RNA, and aligns them so that peptide formation can take place.

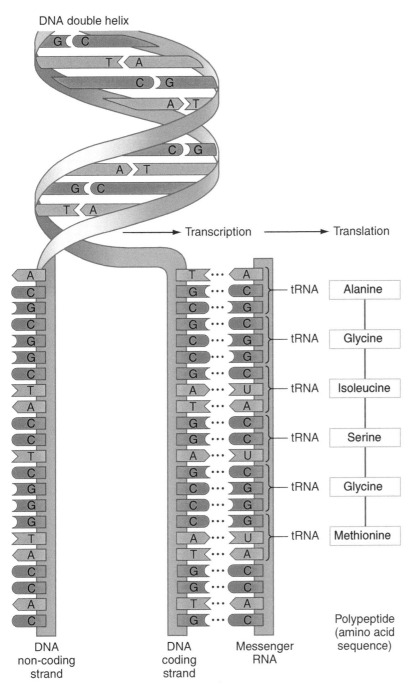

DNA double helix

Transcription ⟶ Translation

tRNA	Alanine
tRNA	Glycine
tRNA	Isoleucine
tRNA	Serine
tRNA	Glycine
tRNA	Methionine

DNA
non-coding
strand

DNA
coding
strand

Messenger
RNA

Polypeptide
(amino acid
sequence)

FIGURE 11.6 Assembly of amino acids into proteins. When the DNA molecule is about to make a protein, a segment of the DNA unzips and a series of nucleotide bases assembles along one of the strands, copying the master sequence.

The assembled string of nucleotide bases, which is slightly different chemically from DNA, is called RNA. As the RNA molecule assembles, it peels off and goes out into the cell. This RNA molecule is called *messenger RNA*, because it carries the message from the DNA molecule to the rest of the cell containing the instructions for building a protein.

The cell contains another kind of RNA molecule, with attachments for amino acids on one end and for nucleotide bases on the other end. This smaller molecule is called *transfer RNA* because it picks up amino acids and transfers them to the proper spot on the messenger RNA, where they add to the growing chain of amino acids that makes up the complete protein.

messenger RNA, it adds the amino acid that it is carrying to the chain of amino acids that has already been built up. When a chain of amino acids has been assembled along the full length of the messenger RNA, the assembly of amino acids into a protein is complete. The newly formed protein then detaches itself from the messenger RNA and moves off into the cell.

By this process, the essential proteins are built up within an organism in accordance with the order of the nucleotide bases in its DNA molecules. In the language of molecular biology, the segments of the DNA molecule are "transcribed" into messenger RNA, and then "translated" into a sequence of amino acids by the transfer RNA. Each DNA segment, controlling the assembly of one protein, is a word; each nucleotide base within a segment is a letter; the order of the letters provides the meaning of the word – that is, the protein to be assembled.

For example, three cytosine rungs in a row (CCC) on the strand of messenger DNA are read by the transfer RNA molecules in the cell as meaning "pick up a proline molecule" (one of the 20 amino acids). This proline molecule is tacked on to the end of a protein being assembled out of amino acids, one amino acid after another in the cell. If the next three rungs on the DNA ladder were, for example, cytosine, adenine, and guanine, in that order (CAG), that would be read by the transfer RNA molecules as meaning "pick up a glycine molecule" (another of the 20 amino acids), and tack it onto the proline that was just picked up. That would be the next step in the manufacture of this particular protein. Eventually the protein – a chain of several hundred amino acids stitched together – is complete. That is how DNA directs the assembly of the proteins that characterize an organism.

Why are *three* rungs of the DNA ladder needed to specify *one* amino acid? The reason is that there are 20 amino acids, but only four kinds of nucleotide bases in DNA. The fifth nucleotide base uracil (U) only occurs in the related molecule RNA, where it replaces thymine (T). Thus, one rung on the DNA ladder would only suffice to pick up four different amino acids; it would only "code" for four amino acids, in the jargon.

If the basic "message unit" were a sequence of two rungs on the DNA ladder, the DNA could "code" for 16 amino acids. That is because there would be four possible choices for a nucleotide base on the first of the two rungs, and for each of these choices there would be four possible choices for the second rung. That would be a total of $4 \times 4 = 16$ separate ways of putting nucleotide base bases on the two rungs. Hence, such a two-rung combination could carry the message to pick up 16 different amino acids.

But 16 combinations are not enough. We need 20 possible combinations for the 20 amino acids. So nature goes on to a three-rung combination. Three rungs offer a total of $4 \times 4 \times 4 = 64$ possible combinations, more than enough. Some of the extra combinations are used for punctuation: "stop" and "start" messages for beginning and ending the assembly of a protein. This is the *genetic code*, first worked out in the 1950s and 1960s (Table 11.1). There is also redundancy built into this genetic code. For example, the combination cytosine–cytosine–adenine (CCA) codes for the amino acid proline, just as the combination cytosine–cytosine–cytosine (CCC) does. For the amino acids arginine, leucine, and serine, there are six different three-rung combinations in the code.

11.4.1 DNA replication

How is the plan for the assembly of the right kind of proteins passed from one generation to the next? How do progeny inherit their characteristics from their parents? The answer lies in a most extraordinary property of the DNA molecule – the ability to make a copy of itself.

Watson and Crick found that DNA consists of two chains, joined together at regular intervals by the pairs of nucleotide bases that run between them like rungs of a ladder. At the middle of each rung of the ladder, between the two nucleotide bases, is a weak spot (called a hydrogen bond), which is easily broken. During the early life of a cell, the two strands remain connected, but when the cell has attained its full growth, and the division into two daughter cells is about to commence, the weak connections running down the middle of the ladder break, and the double strand separates into two single strands.[3] Then a protein within the cell brings nucleotide

---

[3] The assistance of a number of specialized proteins is required for the replication of DNA: to unwind the DNA, to bind the single-stranded areas, to ensure that the proper partner for each nucleotide is selected, and to form the inter-nucleotide bonds in the chain under construction, by eliminating water. These functions are performed by specialized enzymes called DNA polymerases.

Table 11.1 *Genetic code*

		Second base in codon			
		T	C	A	G
First base in codon	T	TTT Phe TTC Phe TTA Leu TTG Leu	TCT Ser TCC Ser TCA Ser TCG Ser	TAT Tyr TAC Tyr TAA *Stop* TAG *Stop*	TGC Cys TGC Cys TGA *Stop* TCG Trp
	C	CTT Leu CTC Leu CTA Leu CTG Leu	CCT Pro CCC Pro CCA Pro CCG Pro	CAT His CAC His CAA Gin CAG Gin	CGT Arg CGC Arg CGA Arg CGG Arg
	A	ATT Ile ATC Ile ATA Ile ATG Met, *Start*	ACT Thr ACC Thr ACA Thr ACG Thr	AAT Asn AAC Asn AAA Lys AAG Lys	AGT Ser AGC Ser AGA Arg AGG Arg
	G	GTT Val GTC Val GTA Val GTG Val	GCT Ala GCC Ala GCA Ala GCG Ala	GAT Asp GAC Asp GAA Glu GAG Glu	GGT Gly GGC Gly GGA Gly GGG Gly

Bases: A, adenine; G, guanine; c, cytosine; T, thymine.
The 20 amino acids specified by each codon

Ala: Alanine
Asp: Aspartic acid
Arg: Arginine
Asn: Asparagine
Cys: Cysteine
Glu: Glutamic acid
Gln: Glutamine
Gly: Glycine
His: Histidine
Ile: Isoleucine

Leu: Leucine
Lys: Lysine
Met: Methionine
Pro: Proline
Phe: Phenylalanine
Ser: Serine
Thr: Threonine
Trp: Tryptophane
Tyr: Tyrosine
Val: Valine

bases into close proximity with the nucleotide bases in each of the now single strands of DNA (Figure 11.7). Each nucleotide base is matched up with its complementary nucleotide base, and two new double-stranded molecules are produced.

There now exist two identical DNA molecules, where formerly there was one. The two DNA molecules separate, and move to opposite corners of the cell; the cell then divides into two daughter cells, each containing one complete set of DNA

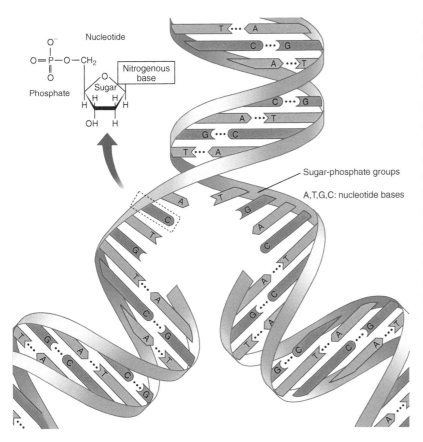

O⁻

O=P—O—CH₂
‖
O

Phosphate

Nucleotide

Nitrogenous base

Sugar
H   H
H   H
OH   H

Sugar-phosphate groups

A,T,G,C: nucleotide bases

FIGURE 11.7 The DNA molecule copies itself. When a cell is about to divide, its DNA molecules unzip, the "ladder" untwists, the rungs break in their midpoints, and the DNA begins to separate into two parallel strands. In the next step, each strand of the unzipped DNA collects new nucleotide bases from the molecules surrounding it in the cell to form a new complete "ladder." The result is two "ladders" that are replicas of the original DNA molecule.

molecules. Thus, each of the daughter cells contains a copy of the complete genetic information that had been in the parent cell. This is the way in which the shape and character of a plant or an animal are transmitted from generation to generation.

In summary, the DNA molecule provides the instructions for the assembly of proteins, and the proteins determine the nature of the organism. Each living organism has its own special set of DNA molecules; no two organisms have the same set unless they are identical twins. However, the basic nucleotide bases and amino acids are the same in every living creature on the face of Earth, whether bacterium, mollusk, or human.

11.4.2 Genes and chromosomes

Before the discovery of DNA and its properties, the mechanism of inheritance was known to reside in the *chromosomes* of the cell nucleus, which can be seen to divide during cell division. Each chromosome was believed to contain a number of *genes* along its length, each gene (or perhaps a group of genes) being concerned with one particular trait of the organism.

Now we know that each chromosome contains a DNA molecule, and that the genes are segments of the DNA molecule, each about 1000 nucleotide bases or "rungs" long, that control (in the manner described above) the assembly of one protein (or major unit of a protein). In addition to segments that control the assembly of proteins, genes also contain non-coding regions (called introns) and regions that regulate the expression of a particular gene.[4] Regions on either side of protein-coding sequences are transcribed into messenger RNA but not translated into proteins.

We say that a gene is a segment of DNA about 1000 nucleotide bases long because proteins (whose assembly the genes control) typically consist of a few hundred amino acids – say, 300 amino acids in round numbers – and 300 amino acids need $3 \times 300 = 900$ nucleotide bases, or 1000 in very round numbers, to code for them.

A *genome* is the complete sequence of DNA in an organism. Bacteria, relatively simple forms of life, have only one DNA molecule, and this molecule has only a few million nucleotide bases. Since 1000 nucleotide bases make a gene (and a protein) a bacterium has only a few thousand genes. A human has 23 pairs of chromosomes, hence 23 pairs of DNA molecules, and each molecule is roughly 100 million nucleotide bases long.

From the above information, a human should have a few million genes, and therefore, a few million different kinds of proteins. Recent results of the Human Genome Project indicate that humans have only about 25 000 to 50 000 kinds of proteins, so it appears that only a small fraction of the DNA in humans actually makes proteins. What does the other 98% do? That is one of the most interesting questions in biology today. Some of the DNA is important for promoter

---

[4] An organism contains all of the genes required for making all the proteins needed to generate that particular organism. Genes can be turned on and off (when more or less of a particular protein is needed). The expression or non-expression of genes requires other proteins that interact with the genes.

sequences, which control when a gene is turned on or off. Some are ancient genes that no longer function, and others may even be genetic parasites that invaded the DNA of human ancestors.

11.5 The origin of life

The newly formed Earth must have been a sterile body. The intense bombardment that accompanied the origin of the planet would surely have melted Earth's surface and vaporized its early oceans. Yet we have geochemical evidence that life may have existed on Earth as early as 3.8 billion years ago. Therefore, either life came here from somewhere else, or life originated on Earth out of non-living chemicals.

Advances in science in recent decades have uncovered facts about the nature of living organisms that lead for the first time to a possible scientific explanation of the origin of life. It now appears likely that the first living creatures on Earth evolved spontaneously out of the inert chemicals that were common on the planet in the early years of its existence.

11.5.1 The Miller–Urey experiment

An early experiment bearing on the origin of life was first performed in 1952 by Stanley Miller. Miller was then a graduate student working on his Ph.D. thesis under Harold Urey. At the suggestion of Urey, Miller mixed together the gases ammonia (NH_3), methane (CH_4), water vapor (H_2O), and hydrogen (H_2), which were thought at that time to have been abundant in Earth's primitive atmosphere. Miller circulated the mixture through an electric discharge (Figure 11.8). At the end of a week, he found that the water contained several types of amino acids.

The key to the formation of biologically important molecules in these experiments is the presence of hydrogen and hydrogen-rich compounds that were thought to have been abundant in Earth's atmosphere early in our planet's history. Some scientists suggest, however, that the early atmosphere consisted of gases such as carbon dioxide (CO_2), water vapor

FIGURE 11.8 The Miller–Urey experiment. Biochemists have manufactured the molecular building blocks of life in the laboratory out of simple ingredients. The first experiment of this kind was performed by Stanley Miller in 1952 working under his thesis director Harold Urey.

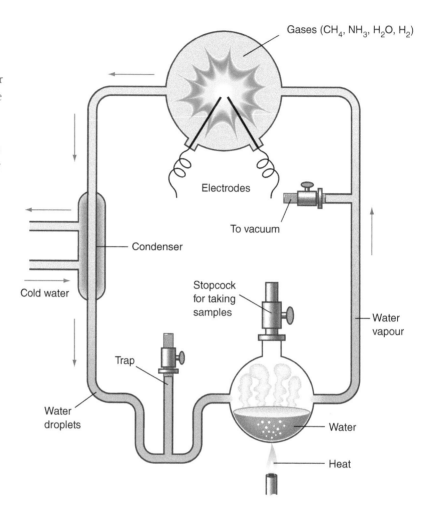

Gases (CH_4, NH_3, H_2O, H_2)

Electrodes

To vacuum

Condenser

Cold water

Stopcock for taking samples

Water vapour

Trap

Water droplets

Water

Heat

(H_2O), and molecular nitrogen (N_2). Experiments using these gases as starting materials also produced some of the basic building blocks of life, but at a lower yield. As long as the experimental setup did not include free oxygen (O_2), however, some organic components were produced.

Under certain circumstances, hydrogen cyanide (HCN) can be formed in the Miller–Urey apparatus. In 1961, biochemist Juan Oro found that amino acids could be made from a mixture of hydrogen cyanide (HCN), ammonia (NH_3), and water. His experiments also produced the nucleotide base adenine. Adenine is also a component of adenosine triphosphate, or ATP, which is a major energy-releasing molecule in cells.

The Miller–Urey experiment seems to suggest that it is possible to create biologically significant molecules, although the particular chemical pathways of life's origin may have been different than those originally investigated by Miller and Urey.[5]

The discovery of primitive bacteria that live in the scalding waters on the rim of submarine hotsprings has generated interest that life could have had its origins in deep volcanic vents. Carbon-rich gases released at deep-sea vents could have reacted with hydrogen sulfide (H_2S) gas to form many of the basic chemicals of life. These deep, hot vents would have provided an environment where the building blocks of life were heated and concentrated, a situation well suited for the creation of more complex molecules.

Some of the molecules made in Miller–Urey type experiments are now known to exist in space. On September 28, 1969, a meteorite fell near Murchison, Australia. Analysis of the meteorite has shown that it contains small quantities of amino acids. If amino acids are able to survive in outer space under extreme conditions, then the early Earth may have acquired some of its amino acids and other organic compounds through asteroid and comet impacts.

Production of more complex molecules seems to require some process that concentrates the basic building blocks of life. One possible method of concentration involves the adsorption of amino acids onto the surfaces of common minerals such as clays. Some scientists think that the crystal structure of clay minerals could have provided a kind of template for the construction of the first complex organic molecules.

No one can say for sure how the origin of life proceeded, but at some stage the combination of nucleotide bases must have produced the first molecule capable of replication, perhaps a simple strand of RNA. With that step, the threshold of life was crossed.

In this early "RNA world," the major life functions of information storage, replication, and protein synthesis were all performed by RNA. The RNA molecules acquired the ability

---

[5] Miller's experiment made several of the amino acids present in proteins, but also a large amount of molecules of no importance in biology. Many "building blocks" important to life have never been prepared in Miller–Urey type experiments, nor in the HCN polymerizations run by Oro.

to catalyze their own replication, and helped to assemble amino acids into proteins that aided RNA replication. Later, as life evolved, a division of labor occurred, with the DNA molecule providing a more stable library of genetic information while proteins took over most of the catalytic functions. RNA continued its role as the go-between in the synthesis of proteins.

11.6 The virus

Is there any direct evidence for the development of life out of non-living molecules? In fact, an entity exists, very common in the world today, which possesses, at the same time, the attributes of a non-living molecule and the attributes of a living organism. This entity is the virus – the smallest and simplest object that can be said to be alive.

The existence of viruses first came to light at the end of the nineteenth century, in the course of a series of experiments designed to reveal the cause of a disease affecting tobacco plants. It was found that the juice pressed from the leaves of infected plants could transmit the infection to other plants. Apparently, the infection was transmitted by bacteria contained in the fluid. But when pressed through a fine filter, which screened out all visible bacteria, the fluid still retained its power of infection.

In 1898 a Dutch botanist, Martinus Beijerinck, suggested that the disease was not caused by bacteria, but by poisonous chemical. Beijerinck called the chemical a *virus*, which is the Latin word for poison. Gradually the suspicion developed that the virus was no ordinary chemical. A variety of experiments suggested that the virus, although too small to be seen under the microscope, possessed the basic attribute of living organisms – the ability to reproduce itself (Figure 11.9).

Still, the evidence for the living virus was indirect; no one had yet seen one in the act of reproduction. But in the years after the Second World War a new instrument was perfected, which provided the biologists with a powerful tool for the study of small organisms. This instrument was the electron microscope. Ordinary microscopes, in which the object under

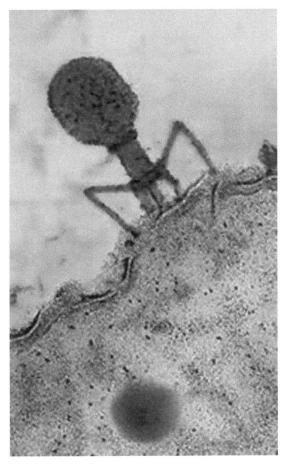

FIGURE 11.9 The virus (with its component parts). The virus lies on the threshold between living matter and inanimate molecules; it is the simplest and smallest living particle, some viruses being only one-millionth of an inch (one tenth of a micrometer) in diameter.

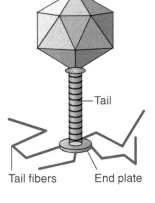

Head (capsule)

Tail

Tail fibers   End plate

study is illuminated by rays of light, are limited to a magnifying power of approximately 2000 times. The smallest bacteria, which are a hundred-thousandth of an inch in size, can just barely be seen in these microscopes. But the electron microscope, which directs a beam of electrons at the object instead of a beam of light, can produce magnifications as high as several hundred thousand times. It is possible to photograph a single protein molecule with these instruments.

Under the electron microscope the virus finally became visible, and all the important details of its structure were revealed. It was found that viruses come in many shapes – round, cylindrical, polyhedral, and with tails (Figure 11.9). They also come in many sizes. The largest is as large as a small bacterium; the smallest, which is a millionth of an inch in diameter, is smaller than many non-living molecules. Viruses bridge the gap in size between the inanimate and animate worlds.

Chemical studies show that viruses contain nucleic acids – the molecular blueprint of life and the means by which every living creature reproduces itself. They also contain a substantial amount of protein, in the form of a protective coat wrapped around the precious, delicate strands of nucleic acid. But they contain very little else. In particular, they have none of the sugar and fat molecules that provide energy for the chemical reactions in other living creatures. They also lack free nucleotide bases and amino acids, out of which all other organisms make proteins and assemble copies of themselves.

How, then, without a source of energy, and without the materials essential for growth and reproduction, do viruses live?

The answer is clearly revealed by the electron microscope. If a solution of virus particles is carefully dried out, the viruses will stick together in a symmetric pattern to form a crystal as geometric – and as lifeless – as a crystal of salt or a diamond; left undisturbed, the crystal remains inert for years (Figure 11.10). But, dissolved again in water, and placed in contact with living cells, the molecules of the crystal spring to life; they fasten to the walls of the cell, dissolve a small opening in the cell wall, and, through the opening, inject their nucleic acid into the cell.

Once inside the cell, the nucleic acid of the virus seizes control, displacing the original DNA of the cell and establishing itself as the master of all further chemical activity. The full molecular resources of the invaded cell – the energy-giving fats and sugars, the amino acids, and the nucleotide bases – are commandeered and employed in the assembly, not of the proteins needed by the invaded cell, but of the proteins needed by the virus.

At the same time, the virus gathers free nucleotide bases in the cell and assembles them, not into copies of the DNA of the invaded cell, but into copies of itself. The virus even secretes an enzyme that breaks down the existing DNA within the cell into its component nucleotide bases in order to have more of these precious units available for making replicas of itself.

FIGURE 11.10 Carefully dried out, a solution of virus particles forms completely inert crystals. Dissolved in water and given access to living cells, the viruses composing the crystal come to life and attack their hosts.

When a large number of protein coats and viral nucleic acids have been assembled, the cell is milked dry. The coats wrap around the virus RNA or DNA molecules to form complete viruses, while the original virus, in a final step, secretes an additional enzyme that dissolves the cell walls. An army of virus particles marches forth, each capable of invading new cells, leaving behind the empty, broken husk of what had been, an hour before, a healthy living cell (Figure 11.11). The operation is simple and effective. It is executed by an organism that is, in the smallest viruses, only 200 atoms wide. The virus is truly the link between life and non-life – the bridge between living and non-living matter.

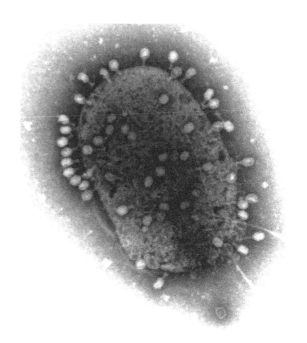

FIGURE 11.11 Viruses gathered around a bacterium. Within an hour, viruses entering this bacterium will consume its chemicals in making replicas of themselves, leaving just the husk of a previously healthy cell.

   Professor M. V. Parthasarathy, Plant Biology/CIMC, Cornell University.

11.7 Evolution

In earlier chapters, we have seen how the basic forces of nature – gravity, electromagnetism, and the nuclear force – acting on the basic building blocks of matter, have led, first to the synthesis of the elements within the interiors of stars; then to the formation of the Sun and planets out of those elements; and finally, on the surface of one of those planets, to the formation of the molecular building blocks of life. Throughout this long history, our viewpoint has been that of the physicist, seeking to understand the essence of the world in terms of a few simple principles. One might call them the laws of physics. These laws are the distillation of all the observations regarding the physical world that have been acquired in thousands of years of human experience.

Now we come to the explanation of the subsequent course of events in the history of life, leading from the first simple organisms to humans. The stars and planets have yielded the secrets of their history to the physicist; the molecular foundations of living organisms are being decoded; but the *complete* organism – even the simplest and most primitive kind – is more complicated than any star, or planet, or giant molecule.

New insights are needed for the understanding of its structure and evolution. A new law must be found.

11.8 Darwin's discovery

The new law was discovered by Charles Darwin (1809–1882) (Figure 11.12). Darwin showed that evolution is the result of a mechanism or "force" in nature, which works on plants and animals slowly, over the course of many generations, to produce changes in their forms. This "force" is not to be found in any textbook of physics, listed alongside the basic forces that control the world of non-living matter; but, nonetheless, it guides the course of evolution and shapes the forms of living creatures – on this planet and on all planets on which life has arisen – as firmly and as surely as gravity controls the stars and the planets.

FIGURE 11.12 Charles Darwin in 1880, two years before his death.

Darwin was led to his discovery by observations of plant and animal life carried out between 1831 and 1836, during a voyage around the world on HMS *Beagle*, a British navy vessel assigned to surveying and mapping duties in the Southern Hemisphere. He had sailed on the *Beagle* as a naturalist, serving without pay and collecting specimens during a journey that carried him around most of the South American continent, to Australia, New Zealand, Africa, and many Atlantic and Pacific islands.

From 1832 to 1835 the ship sailed up and down the coast of South America, and on several occasions during that time Darwin went ashore for long overland journeys of exploration. During these trips ashore, Darwin unearthed the evidence that first turned his mind toward evolution. He came upon beds of fossils containing the skeletons of animals that had once roamed the Argentine Pampas but had been extinct for tens of thousands of years.

None of these extinct creatures was the same as any animal now alive on the South American continent; and yet some of them bore a surprising resemblance to existing species. Darwin thought about this "wonderful relationship . . . between the dead and the living;" was it possible that all the animals living on Earth were directly descended from vanished

FIGURE 11.13 Darwin's finches. These birds occupy neighboring islands in the Galapagos Islands. Their beaks are well adapted for the types of seeds and grubs available on the specific islands on which they live. This diverse group of birds descended from a few finches that arrived on the islands from the South American mainland.

Campell, Reece, Mitchell (1999) *Biology*, 5th edition. Benjamin/ Cummings@Menlo Park, CA

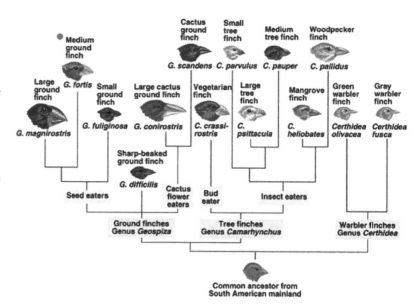

species? Could it be that the passage of vast amounts of time had, in some way, worked the changes between those ancient animals and their modern descendants?

Other facts impressed Darwin later in the voyage. The *Beagle* stopped for a month at the islands of the Galapagos Archipelago, situated on the Equator 500 miles (800 kilometers) west of the coast of South America. During the visit Darwin noticed a "most remarkable feature" of these islands: although they were close to one another, and the climate and soil were similar on all, different kinds of plants and animals lived on them. Some of the islands even contained plants or animals that were to be found on those islands and on no others (Figure 11.13).

Darwin puzzled over the curious nature of these variations in species; if all forms of life on Earth had been placed here by separate acts of creation, why was the creative force so prodigal in bestowing separate species on each island in the Galapagos?

Another observation led Darwin to the answer. He had also noticed that most of the animals on the islands resembled animals that were peculiar to the neighboring South American continent, and were not to be found in other parts of the world. The significance of this fact did not occur to Darwin until his return from the voyage of the *Beagle* in 1836. Then an explanation occurred to him: a long time ago plants, insects, birds, reptiles, and mammals, carried by currents of air and wind, or floating on driftwood, must have reached the Archipelago from the adjacent coast. Isolated from the mainland, they evolved into forms that came to differ more and more, in the course of time, from those of their mainland cousins. Moreover, when the migrant plants and animals first arrived at the Archipelago and became established, they were identical on every island; but gradually, because the islands were isolated and cross-breeding between islands rarely occurred, distinct lines of evolution developed on each. In this way the separate islands acquired their characteristic flora and fauna.

Darwin's reasoning implied that the forms of life could change and evolve with the passage of many generations. By 1838 he was convinced that "such facts as these ... could only be explained on the supposition that species gradually become modified." He was convinced of the truth of evolution.

Throughout the years following the voyage of the *Beagle*, Darwin's mind turned on the problem of a cause for evolution. Gradually the outlines of a new theory emerged. By the end of 1838 the new law of nature was clearly formulated in his notebooks. According to Darwin, he found the catalyzing element for his theory in an essay by Thomas Malthus, a Scottish clergyman who stressed the tendency of populations to increase rapidly in numbers without limit and the harsh struggle for survival that resulted. In Darwin's autobiography, he wrote:

In October 1838, that is, fifteen months after I had begun my systematic enquiry, I happened to read for amusement "Malthus on Population", and being well prepared to appreciate the struggle for existence which everywhere goes on, from long-continued observation of the habits of animals and plants, it at once struck me that under these circumstances favourable variations would tend to be preserved, and unfavourable ones to be destroyed. The results of this would be the formation of new species. Here, then, I had at last got a theory by which to work.

Yet he did not announce it to the world immediately; he knew that his belief in evolution would make him an unpopular figure. At last, in November 1859, Darwin's theory appeared in print with the title of *The Origin of Species by*

Means of Natural Selection or The Preservation of Favored Races in the Struggle for Survival.

11.9 Natural selection

The argument set forth in *The Origin of Species* was beautifully simple and clear; its validity should have been apparent to everyone. Darwin began with an almost self-evident set of observations on the nature of life: All living things reproduce themselves; reproduction is the essence of life; *but the process of reproduction is never perfect.* The offspring in each generation are not exact copies of their parents; brothers and sisters differ from one another; no two individuals in the world are exactly alike, except for identical twins at the moment of birth.

Darwin asserted that these small variations are critically important; for, in the struggle for existence, the creature that is distinguished from its brethren by a special trait, giving it an advantage in the competition for food, or in the struggle against the rigors of the climate, or in the fight against the natural enemies of its species – that creature is the one most likely to survive, to reach maturity, *and to reproduce its kind.* Some of the offspring of the favored individual will inherit the advantageous characteristic; a few will possess it to a greater degree than the parent. These individuals are even more likely to survive and to produce offspring. Thus, through successive generations, the advantageous trait appears with ever-increasing strength in the descendants of the individual who first possessed it.

Not only does the trait become more pronounced in each individual with the passage of successive generations, but the number of individual animals possessing it also increases. These favored individuals tend to have slightly larger families than the average, because they and their offspring have a greater chance of survival to reproductive age; in each generation they leave behind a greater number of offspring than their less-favored neighbors; their descendants multiply more rapidly than the rest of the population, and in the course of many generations, their progeny replace the progeny of the animals that lack the desirable trait.

In *The Origin of Species*, Darwin gave this process the name by which it is known today: "This principle of preservation or the survival of the fittest, I have called *Natural Selection*."

Through the action of natural selection, a favorable trait that first appeared as an accidental variation in a single individual, will, with the passage of sufficient time, become a pronounced characteristic of the entire species. So the deer become fleet of foot, for the deer that ran fastest in each generation usually escaped their predators and lived to produce the greatest number of progeny for the next generation. So did humans become more intelligent, for superior intelligence was of premium value: the intelligent and resourceful hunter was the one most likely to secure food. Thus developed the human brain; thus too, in response to other pressures and opportunities in their environments, developed the trunk of the elephant and the neck of the giraffe.

Of course, the incorporation of one new trait does not create an entirely new animal. But if we count all the births that occur to a single species over the face of the globe in one year, an enormous number of variations will appear in this multitude of young creatures. On all these variations, the same process of selection works steadily, preserving for future generations the new traits that give advantages to the individuals of a species, and eliminating those that do not. The changes may be imperceptible from one generation to the next, but over the course of many generations the accumulation of many favorable variations, each slight in itself, completely transforms the organism.

Natural selection molds the forms of life. Under its continuing action the shapes of animals change with time; old species disappear in response to changing conditions, and new ones arise. Few of the species of animals that roamed Earth 10 million years ago still exist today, and few of those existing today will survive 10 million years hence. To quote again from *The Origin of Species*: " . . . not one living species will transmit its unaltered likeness to a distant futurity." But natural selection works its effects subtly. Its influence is not felt in one individual or in their immediate descendants.

A thousand generations may elapse before a change becomes noticeable; in humans that amounts to twenty thousand years. Yet, ever since Earth's age was accurately measured, we have known that enough time is available. Our planet has existed for billions of years; that is the secret strength of Darwin's theory. "We have almost unlimited time," he wrote in 1858, in explaining how the slightest variations in the form of an animal can grow, through their effect on the probability of producing progeny, until, after the passage of "millions on millions of generations," great changes are effected. And in *The Origin of Species*:

The mind cannot grasp the full meaning of the term of even a million years; it cannot add up and perceive the full effects of many slight variations, accumulated during an almost infinite number of generations ... We see nothing of these slow changes in progress, until the hand of time has marked the lapse of ages, and then ... we see only that the forms of life are now different from what they formerly were.

The arguments over Darwin's views gradually subsided through the decade of the 1860s. They erupted again when *The Descent of Man* was published in 1871. In this book, Darwin presented his views on human origins and history, confirming the darkest suspicions of his critics by setting forth evidence linking humans and apes to a common ancestor. But by Darwin's death in 1882, his theories were widely accepted in the scientific world, and made a substantial impact on human thinking.

11.10 DNA, mutations, and evolution

Throughout the years in which Darwin's ideas were winning increasing acceptance, one point remained obscure. What was the origin of the variations from one individual to another that provided the raw material for natural selection? Regarding these variations, which played so essential a role in his theory, Darwin could only say helplessly, "We are profoundly ignorant of the cause of each slight variation or individual difference ... [They] seem to us in our ignorance to arise spontaneously." The ignorance was not fully dispelled until 1953, nearly a century after the publication of *The Origin of Species*, when it became clear how the basic characteristics of the individual are passed from generation to generation. These characteristics reside in the molecule called DNA, which is found in the cells of every living organism on Earth. As we have noted, no two individuals in the world, except identical twins, have the same sequence of nucleotide bases in their DNA. This sequence determines which proteins will be assembled in the cells of the body; and the proteins, in turn, control the body chemistry and all the traits of the individual. Thus, the DNA in every creature contains the master plan for that creature.

We now know that occasionally some of the nucleotide bases in the DNA molecule are damaged, altered, or removed entirely from the molecule, so that the master plan is changed. The damage or alteration may affect only one nucleotide base in the long chain – a chain that may, in human cells, stretch out over a billion nucleotide bases. Nevertheless, the change in a single nucleotide base may be critically important, for the proteins in the cell are assembled out of amino acids in a sequence that follows the order of the nucleotide bases in the DNA. Damage to one of these nucleotide bases, or replacement of one nucleotide base by another of a different type, will lead to the assembly of a slightly different protein, in which, at one point along the chain of amino acids making up the protein, the wrong kind of amino acid is located.

When a modified DNA and the modified protein produced by it are situated in an ordinary cell of the body, the abnormal cell is soon replaced by the growth of new cells and the effect of the change in the DNA molecule quickly disappears. However, in one type of cell in the body a change in the sequence of nucleotide bases in the DNA molecule may have permanent and serious consequences. This is the germ cell – sperm in the male and ovum in the female. Like all other cells, the germ cell contains its set of DNA molecules with the master plan for the development of the individual.

When the sperm and ovum unite to form a fertilized egg, every organ in the body of the mature individual subsequently develops out of the egg by repeated cell division, following a new, joint master plan provided by the combination of the DNA molecules from the sperm and the ovum participating in the union. If the DNA in one of these two germ cells has been damaged or altered, the effect of the change will appear in every cell of the body of the new individual. Moreover, it will be transmitted to the offspring of this individual in the next generation, and in every generation thereafter. All the descendants of that individual, down through the corridors of time, will bear the mark of the change in the sequence of nucleotide bases in the ancestral DNA.

A modification of the nucleotide bases in the DNA of a germ cell is called a *mutation*. Mutations are changes in the body chemistry of the individual that are transmitted to its progeny. They are the inheritable variations that form the basis of Darwin's theory of evolution.

Some mutations change the chemistry of the body in a way that improves the individual's reproductive success; these are called *favorable* mutations. The individuals possessing a favorable mutation are always the ones most likely to propagate their species; from generation to generation their number steadily increases, and in the course of many generations the favorable mutations spread throughout the population. Mutations may also be *unfavorable*, diminishing reproductive success. Over the course of generations, these unfavorable mutations are gradually weeded out of the population. That is the way in which evolution works: By the pruning away of unfavorable mutations simultaneously with the strengthening of favorable ones.

What is the cause of mutations? What can damage or modify the sequence of nucleotide bases in the germ cells of an organism?

One cause of mutations lies in the cell itself. Occasionally an error – an insertion or deletion of nucleotide bases or the mistaken duplication of an entire gene – occurs in the copying process by which the DNA molecule reproduces itself. At the start of the process of reproduction, just prior to the division of a cell into two daughter cells, the double-stranded DNA in the parent cell unwinds and separates into two single strands. With the help of specific proteins, each strand gathers new nucleotide bases from the pool of nucleotide bases within the cell to form a new, double-stranded DNA duplicating the original. It is at this point that the copy error may occur. For example, one of the newly added nucleotide bases may be of the wrong kind; that is, it fails to match its counterpart in the existing strand. As a result, when the assembly of the two DNA molecules has been completed, the sequence of nucleotide bases in one of the daughter

DNAs differs from the sequence of nucleotide bases in the parent DNA. That daughter DNA has suffered a mutation.

Copy errors are not the only source of mutations. Cosmic rays – energetic particles produced in distant regions of the Universe – bombard Earth from all directions. Occasionally, either a cosmic ray or radiation produced by artificial sources will pass through a germ cell, even if the cell is buried deeply in the body, and will disrupt the normal sequence of nucleotide bases in its DNA, producing a mutation. According to Darwin's explanation for the history of life, these mutations are the raw materials of evolutionary change.

11.11 Summary

By the end of this chapter you should be familiar with the following concepts and topics:

The origin and evolution of life
What is life?
The basic building blocks of life
   Carbon compounds
   Proteins
The genetic code for life: DNA
   DNA replication: How to pass the code on to future
      generations
   Genes and chromosomes
The origin of life
Charles Darwin: The theory of evolution
   Natural selection
   Mutations

Questions

11.1. What were the initial ingredients and final products of the Miller–Urey experiment?

11.2. What are the basic building blocks of life, both on the elemental and molecular level?

11.3. How does the order of nucleotides on the DNA molecule represent a genetic code for the amino acids that build proteins?

11.4. How is life defined? Discuss life defined as "reproduction with mistakes."

11.5. What are mutations?

11.6. How does natural selection increase the population of organisms with certain "desirable" traits?

11.7. What is Darwinian "fitness"?

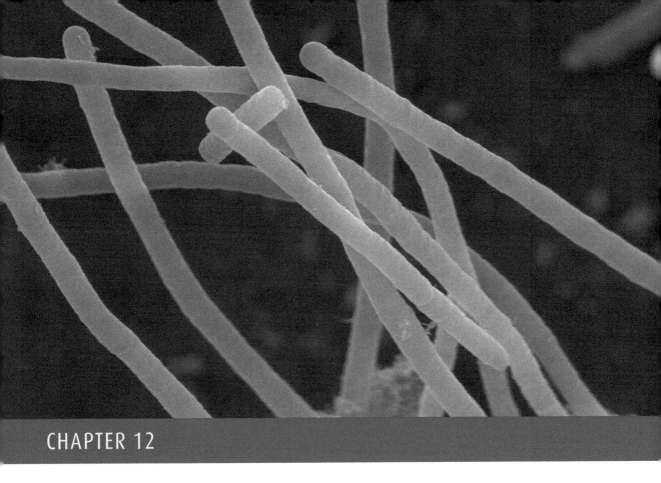

The early history of life on Earth

From single cells to complex organisms

Life originated on Earth with the appearance of the first self-replicating molecules. These simple life forms would have multiplied rapidly, and random mutations caused by copy errors or cosmic radiation produced a spectrum of variations. Those variations that reproduced most rapidly grew in numbers. Soon Darwin's law began to select organisms with innovations that enhanced their reproductive success. Then sexual reproduction arose, in which genetic material from two parents was shuffled to produce new combinations in the offspring. This greatly extended the possibilities for variation and rapidly accelerated the pace of evolution. From simple beginnings, life evolved into a remarkable diversity of forms both in the seas and on land.

The tree of life maps the relationships amongst all the organisms living today, and shows that all were derived from a single common ancestor. Beginning with the formation of the cell, we will now follow the development of life from its simplest, earliest forms, through to the explosion of complex multi-cellular organisms. We will see how the early forms of life got the energy they needed to function and reproduce, and how the process of photosynthesis may have come about. Following evidence from the fossil record, we will learn how and when multi-cellular organisms developed, and why they had a distinct advantage over their single-celled relatives. Moving then to the Cambrian explosion, we will follow the development of life from simple hard-bodied organisms, to fishes, to the momentous development when life invaded the land, culminating in the evolution of complex creatures and eventually humans.

12.1 The cell

One of the first major steps in the evolution of life on Earth was the appearance of the first one-celled organism. The evolution of cells requires a membrane to surround and protect the complex molecules of life. David Deamer and colleagues at the University of California at Santa Cruz have produced small cell-like structures from molecules called *lipids*. One end of a lipid molecule has an affinity for water, whereas the other end is repelled by water molecules. When placed in water, lipid molecules will form a surface film that tends to curl up to produce tiny spheres (Figure 12.1). The hollow and porous lipid spheres can enclose large molecules such as RNA, DNA, and proteins while allowing smaller molecules to pass through.

A strand of DNA trapped inside such a proto-cellular sphere would have an advantage over free DNA molecules because, first, the DNA molecule would be protected from the vagaries of the environment, and second, the concentration of useful molecules would become higher within the cells, since small molecules like water or simple carbohydrates could pass in and out through the thin, porous membrane, while larger molecules, such as complex carbohydrates, nucleotides, or

Previous page: *Planktothrix agardhii* (strain CYA34), filamentous cyanobacteria. Patricia Sánchez-Baracaldo, School of Biological Sciences, University of Bristol

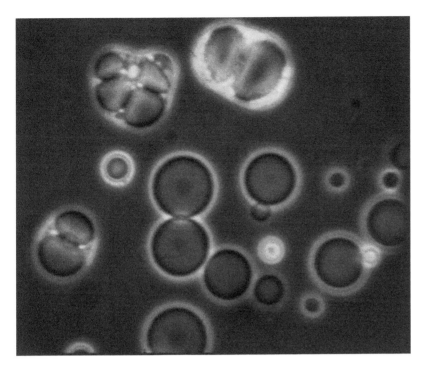

FIGURE 12.1 Spherical bodies from 10 to 50 micrometers in diameter formed in the laboratory by David Deamer and colleagues using lipid molecules.

David Deamer, Jack Baskin School of Engineering, Univeristy of California at Santa Cruz.

amino acids once formed, would be trapped inside. Eventually, nucleic acid sequences evolved that produced proteins able to direct the construction of cell membranes. Through natural selection, these primitive cells would soon become the dominant form of life on the planet.

The oldest rocks found on Earth are about 4 billion years old. The oldest fossil evidence of life is found in rocks of the Warrawoona Formation in northwestern Australia that are about 3.5 billion years old. These earliest fossils resemble in their form modern bacteria and cyanobacteria (photosynthetic one-celled organisms also called blue-green algae) (Figure 12.2).[1] Bacteria and cyanobacteria are primitive cells that lack a nucleus; their DNA is dispersed within the cell, and they are therefore called *prokaryotes* (which means "pre" nucleus). Prokaryotic cells are simple forms of life, but are nonetheless quite complex chemical factories; the smallest bacteria can make about 500 different kinds of proteins.

---

[1] Recently, however, the biological origin of these structures has been questioned. They have been reinterpreted by some scientists as chains of crystals formed in ancient hot springs.

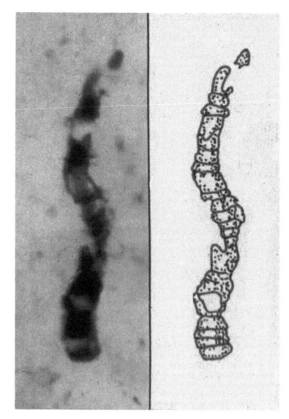

FIGURE I2.2 Microstructure (about 70 micrometers long) interpreted as a fossil cyanobacterium from Western Australia. This controversial fossil is about 3.5 billion years old, and may be the oldest known fossil of cyanobacteria. The alternative interpretation is that it represents a chain of mineral crystals.

J. William Schopf, Department of Earth and Space Sciences, The University of California at Los Angeles

The first self-replicating molecules probably found abundant raw materials with which to build copies of themselves. The primordial waters might have been teeming with organic building blocks, such as amino acids, nucleotide bases, carbohydrates, and lipids. But as the population of self-replicators increased, they would have rapidly depleted the available organic molecules.

The earliest forms of one-celled life most likely derived their energy from metabolic processes like *fermentation* and *methanogenesis*, using organic molecules such as sugars as foodstuff to produce energy, and creating as waste products methane, carbon dioxide, and other organic chemicals. Fermentation utilizes relatively simple enzymes, and requires little in the way of complex cellular chemical machinery. There must, however, be a continuous supply of sugars. On the early Earth, the source of sugars could have been the conversion of atmospheric gases into organic compounds by lightning or ultraviolet radiation. But as the population of ancient micro-organisms approached the limits of the supply of energy-rich molecules such as sugars, severe competition developed for these resources. There was, however, another available source of energy – in an unlimited supply. That source of energy is sunlight.

12.2 Photosynthesis

An organic molecule eventually appeared that had the ability to trap efficiently and store temporarily the energy of sunlight. This molecule absorbs the light energy and is raised to a higher level of excitation. The molecule is called *chlorophyll*. In its structure, chlorophyll resembles an antenna that captures solar energy and begins to vibrate. In cells, chlorophyll is linked with proteins in an intricate assembly called a *photosystem*, a chain of molecules that can pass the solar energy along like a series of vibrating tuning forks. The energy can then be used inside the cell to take apart simple compounds, such as water and carbon dioxide, and put the resulting molecules back together to form more complex organic molecules such as sugars. This process is called *photosynthesis*.

How did molecules like chlorophyll originate? The answer must be that one of the complex molecules within a cell (most likely a pigment that could absorb light and ultraviolet radiation) acquired the ability to store solar energy by a chemical accident. The one-celled organism that possessed chlorophyll could photosynthesize energy-rich molecules from abundant water and carbon dioxide through the chemical reaction:

$$CO_2 + H_2O + energy \rightarrow C_6H_{12}O_6 + O_2,$$

which produces sugar, commonly with oxygen as a by-product. Such an organism had a great advantage over others that could not get their energy directly from sunlight.

The earliest evidence for some kind of photosynthesis in the fossil record is found in rocks that are about 3.8 billion years old. The evidence comes not in the form of fossil organisms, but in the traces of carbon found in these ancient rocks. When this carbon is analyzed for its isotopic composition, we find that it is composed of two isotopes, carbon-12 and carbon-13. *Isotopes* are varieties of the same element with different numbers of neutrons in their nuclei. Carbon atoms always have 6 protons in the nucleus, but can have 6, 7, or even 8 neutrons in the nucleus, giving the isotopes carbon-12, carbon-13, and carbon-14. (Carbon-14 is an unstable isotope that decays relatively rapidly into nitrogen; here we are interested only in the stable isotopes of carbon: carbon-12 and carbon-13.)

In the atmosphere, the ratio of carbon-13 to carbon-12 is relatively constant, but when organisms incorporate carbon into their bodies in the process of photosynthesis, they find that chemical reactions involving carbon-12 are more rapid than those involving carbon-13. Because of this, relatively more carbon-12 is incorporated into the organism than carbon-13. When the cells die, their organic debris is relatively enriched in carbon-12 compared to the carbon-13/carbon-12 ratio in the atmosphere. This enrichment in carbon-12 is found in ancient sediments, and it shows that life may have been widespread as long as 3.8 billion years ago.

In rocks as old as 3.5 billion years, geologists find layered moundlike structures called *stromatolites*, which were formed by colonies of micro-organisms including cyanobacteria. Stromatolites are among the oldest organically produced features found on Earth. Similar mounds built by colonies of microbes are still forming today. They have remained largely unchanged since their appearance on Earth more than 3.5 billion years ago (Figure 12.3).

12.2.1 The growth of atmospheric oxygen

Oxygenic photosynthesis also had a very important effect on the composition of Earth's atmosphere. Cyanobacteria take in carbon dioxide and water and give off oxygen as a waste product. (Some bacteria use hydrogen sulfide (H_2S) instead of water in photosynthesis, and thus do not produce oxygen.) Thus, once oxygenic photosynthesis evolved, oxygen would gradually have begun to accumulate in the atmosphere. At first, the oxygen would be used up mainly in oxidizing iron dissolved in the oceans; later it would begin to build up as dissolved oxygen in the ocean waters, and eventually as

(a)

(b)

free oxygen in the atmosphere. It is interesting to note that vast deposits of iron oxide were precipitated from the early oceans mostly from about 2.5 to 2.0 billion years ago (Figure 12.4). These thick *banded iron formations* were the result of reactions between soluble iron (FeO) and oxygen dissolved in ocean waters. Oxidized iron (Fe_2O_3) is relatively insoluble and quickly settles out of the water. In today's oxygen-rich atmosphere, this oxidation of iron takes place long before the iron reaches the sea, and very little iron is dissolved in seawater.

Beginning about 1.8 billion years ago, we find a new type of sedimentary rock called *redbeds*. Redbeds are red-colored sandstones and mudstones that were deposited in river valleys and floodplains. The red color is a result of iron oxide (or rust) in the form of the mineral hematite (Fe_2O_3). The hematite comes

FIGURE 12.3 Living and fossil stromatolites.
(a) Living forms at Shark Bay, Western Australia. Courtesy of Paul Harrison (b) Fossil stromatolites exposed on the eroded surface of ancient limestone in the northeastern United States.
J. Deuel, Petrified Sea Gardens, Inc. (www.petrifiedseagardens.org)

FIGURE 12.4 Ancient banded iron formation (circled pocket knife at left for scale).

Gordon Gross and Paul Hoffman, Harvard University

from oxidation of iron through reactions with oxygen in the atmosphere. Redbed sediments rich in hematite are evidence that a substantial concentration of oxygen, at least 1% of present levels, existed in the atmosphere by about 2 billion years ago.

12.3 Major steps in the early evolution of life

12.3.1 The Tree of Life

Most organisms on Earth today are still simple prokaryotic cells, i.e., cells without a nucleus. Biologists have used the molecular makeup of organisms to construct a *Tree of Life* showing the relationships among all of the organisms living today (Figure 12.5). Carl Woese of the University of Illinois noted that changes in an organism's DNA alter the sequence of nucleotides in certain RNA molecules within the cell. Since closely related organisms have similar DNA, their RNA is also similar. More distantly related organisms have more differences in their RNA. Thus, a comparison of the RNA from members of various groups of organisms gives an indication of how closely they are related.

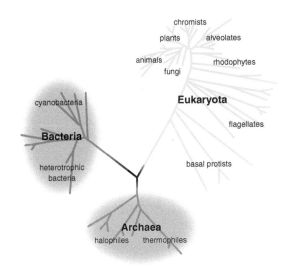

FIGURE 12.5 The Tree of Life showing the evolutionary relations among organisms living today. The tree was constructed using similarities in the DNA of the various organisms. All of these organisms are microscopic except for some members of the plant, fungi, and animal groups within the eukaryotes.

The Tree of Life constructed in this way has three major branches: (1) Bacteria and cyanobacteria; (2) Archaea – prokaryotes that today can be found mainly in extreme environments; and (3) Eukaryota – cells with a proper nucleus. These were all derived from a single common ancestor.

Some studies suggest that the common ancestor of archaea and bacteria were hyper-thermophiles – organisms that lived in environments where temperatures reached 80 °C and higher. If true, the early Earth may have had an atmosphere rich in gases like carbon dioxide and methane creating a hot greenhouse climate.

12.3.2 The first cells with nuclei

Eukaryotic cells are the building blocks of all complex multi-cellular organisms – plant and animal – on Earth today. They represent a major innovation in the history of life. The earliest clear evidence of cells with a nucleus, within which the DNA of the cell is concentrated, occurs in rocks about 2.1 billion years old (Figure 12.6). In these cells, the nucleus acts as a control center where the instructions in the DNA molecule can be read, thus providing centralized direction for the construction of essential proteins and other molecules. Furthermore, the neatly packaged DNA can also carry a fuller set of blueprints, allowing for more complex and efficient cellular functions.

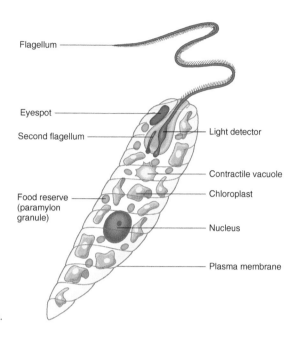

FIGURE 12.6 A eukaryotic cell (cell with a nucleus), showing the nucleus containing DNA. The typical size for this kind of cell is between 50 and 100 micrometers.

How did the first eukaryotic cells originate? Biologist Lynn Margulis of the University of Massachusetts, Amherst, has shown that the eukaryotic cell evolved by the merging of several types of simpler prokaryotic cells. This idea, called *symbiogenesis* proposes that some cells that were engulfed and became trapped within other cells continued to function as individual units. In some cases, the new cell formed by the combination of the two original cells was an improvement in some way over the separate cells. The ingested cells eventually became functioning parts, called *organelles*, of the new eukaryotic cell (Figure 12.7).

This idea is supported by the fact that some organelles in eukaryotic cells have their own separate DNA; when the cell reproduces, the organelles, especially a type called the *mitochondrion*, also divide and reproduce their own DNA (Figure 12.8). The mitochondria are essential in the utilization of oxygen to metabolize foodstuff in eukaryotic cells, as described in the next section. These might be considered the first "animal cells" or *protozoa*. Host cells that ingested cyanobacteria gained the ability to photosynthesize. Eventually, some of these cyanobacteria became the *chloroplasts* (the organelles where photosynthesis takes place) of the first true plant cells (Figure 12.9).

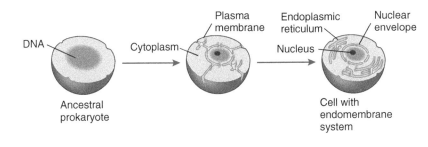

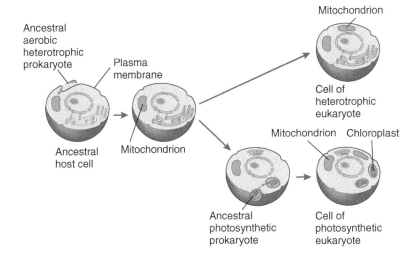

FIGURE 12.7 The first eukaryotic cell was created when one cell engulfed others, which eventually became organelles such as mitochondria and chloroplasts.

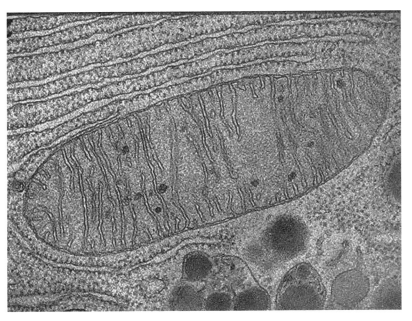

FIGURE 12.8 Electron micrograph of a typical mitochondrion in a eukaryotic cell (length is about 5 micrometers).

Reproduced from "Plant Cell Biology on DVD", www.plantcellbiologyonDVD.com, by B E S Gunning 2007

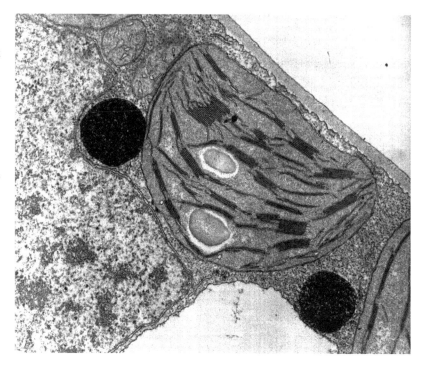

FIGURE 12.9 Electron micrograph of a section through a chloroplast (about 5 micrometers long). The layered structures where chlorophyll resides are similar to those seen in cyanobacteria.

Reproduced from "Plant Cell Biology on DVD", www.plantcellbiologyonDVD.com, by B E S Gunning 2007

12.3.3 Utilization of oxygen

The early prokaryotic cells were capable of building proteins and nucleic acids from more simple precursor molecules. Some organisms such as plants can make all of their organic compounds from a few simple molecules (CO_2, H_2O) that they obtain directly from the environment. But before these basic ingredients can be built into larger molecules, a number of chemical reactions must go on within the cell. This requires a source of energy, and the machinery to convert it into a usable form in the cell. This is where a substance called *adenosine triphosphate* (or ATP) comes into play – it is the energy currency of a cell.

ATP is the most important of a group of molecules that have so-called high-energy bonds between groups of atoms. When these bonds are broken, the energy that is released can be utilized by the cell. ATP can carry energy from one part of a cell to another; this energy can then be used to break apart molecules or to construct large molecules from their constituent parts.

What is the actual chemical mechanism by which cells get energy from food? The simplest kind, used by most prokaryotic cells, is a familiar one – *fermentation*, which involves splitting of sugar molecules to release energy. This energy can then be used in the production of ATP. These reactions need not involve the use of oxygen. However, if oxygen is introduced, then the reactions within the cell can produce much more of the energy-rich ATP. For example, the oxygen utilizing process, called *respiration*, can produce about 20 times as much ATP as fermentation, from the same amount of sugar. The chemical reaction that describes respiration:

$$C_6H_{12}O_6 + O_2 \rightarrow CO_2 + H_2O + \text{energy},$$

is just the reverse of the oxygenic photosynthesis reaction described earlier. Clearly, cells that could utilize oxygen in their metabolism as the free oxygen was building up in the environment could extract much more energy from the same amount of basic foodstuffs.

But oxygen is a dangerous substance to keep within a cell – oxygen can oxidize or "burn up" chemicals essential for the survival of the cell. The utilization of oxygen required the development of a series of enzymes that could safely handle the potentially dangerous oxygen within the cell. In modern eukaryotic cells, the mitochondria (Figure 12.8) act like tiny power plants, where fuel (in the form of organic substances such as sugars) is "burned" (oxidized) to produce energy. As discussed earlier, these mitochondria were originally simple prokaryotic cells that first developed the art of oxygen utilization. These earliest *aerobic* bacteria, the most probable ancestors of cellular mitochondria, could have appeared earlier than 2 billion years ago.

The development of an oxygen-utilizing metabolism was an important step in the history of life. It allowed cells to store, transport, and release great amounts of energy from relatively small amounts of foodstuffs. These highly energetic eukaryotic cells eventually gave rise to all modern multicellular life.

12.3.4 The origins of sex

Organisms that reproduce sexually first appeared in the history of life sometime prior to 1 billion years ago. Sexual reproduction, in which two cells each provide half of the DNA to the daughter cell, allows for more variation than asexual reproduction. When cells simply split in two, forming identical offspring, only chance variations can occur – the result of a cosmic ray, or errors made in the copying process. Such mistakes may happen only once in several hundred replications. (Some bacteria can perform a kind of "sex" by transferring DNA between individuals and even among different species.)

Sexual reproduction, by contrast, involves recombination and shuffling of the genetic material (DNA) from two parent cells, and thus greatly extends the possibilities for variation in a population of cells. The fossil record shows that once sexual reproduction comes into play, the pace of evolution accelerates.

With the evolution of sex, however, comes the question of death. As long as cells reproduced by simple cell division, one parent cell became two daughter cells, and no cell need perish in the process. With sexual reproduction, each parent cell creates an immature cell called a *gamete* with only half the necessary DNA for the creation of a new organism. Therefore, two gametes must combine to produce a complete organism. Once the parent cells have created the gametes, however, their reproductive function is over. Although there would be strong selection against any mutation that could harm a cell prior to reproduction, there would be little or no selection against mutations that lead to cell damage after reproduction. As these mutations accumulate, the parent cells work less and less well. The accumulation of these molecular errors causes the cells to age and eventually die.

12.3.5 Multi-celled organisms

The next major step in evolution was the appearance of organisms with many cells. How did these new organisms develop? The answer must be that some cells acquired, through a chance mutation, the ability to stick together, in simple colonies at first, but later as specialized cells in true multi-cellular organisms. Multi-celled life has distinct advantages. Many-celled organisms can be larger and more resistant to possible damage by predators or swift ocean currents. Cells specialized for certain activities such as digestion, movement,

(a)   (b)   (c)

and sensation can cooperate efficiently for the advantage of the whole organism.

Evidence for abundant multi-celled animals first appears in the fossil record around 1 billion years ago, but some possible multi-cellular fossils occur in rocks as old as 1.6 to 1.4 billion years. Despite what must have been a proliferation of these multi-cellular life forms in the sea, the fossil record of this early life is still rather sparse. This is because these animals were flat and soft-bodied, and thus seldom preserved. There were, as yet, no organisms possessing internal or external hard parts.

From about 630 to 540 million years ago, a characteristic group of multi-cellular animals populated the seas. These unusual organisms are called *Ediacarans* after a locality called Ediacara in Australia where they are found as fossils. First discovered in southwest Africa, they have been found in similar age rocks all over the world. The base of the beds bearing Ediacaran fossils is dated at about 630 million years ago, and rests on older glacial deposits of a geographically widespread ice age (the last of the Snowball Earth episodes). The Ediacarans (Figure 12.10) may be ancestors of more

FIGURE 12.10 Representatives of the Ediacara fauna of 630 to 540 million years ago. The organism on the left is a segmented animal, the others are flat, disk-shaped forms.

© 1994–2002 The University of California Museum of Paleontology, Berkeley, & the Regents of the University of California

advanced multi-cellular organisms that appeared about 540 million years ago, or they may represent an early experimental form of many-celled organisms that reached an evolutionary dead end, and suffered a mass extinction.

12.4 The Cambrian explosion: The appearance of hard-bodied organisms

About 540 million years ago, at the beginning of the Cambrian Period of geological time, the fossil record exploded into a profusion of new organisms with a variety of body plans. During a period of a few tens of millions of years, animals recognizable as arthropods (relatives of insects and crustaceans), mollusks, and even primitive vertebrates appeared for the first time. Many of these organisms possessed external skeletons – hard parts made of the minerals calcium carbonate and calcium phosphate.

Why did so many new animal types evolve at this time? And why did many of these organisms suddenly acquire the ability to secrete hard shells? One theory for the so-called *Cambrian explosion* of life points to increased predation and competition in the Cambrian seas as a driver for rapid animal evolution. For the first time predators made their appearance. The establishment of a food chain may have been partly driven by the independent evolution of vision in several groups of animals. It may be no coincidence that the Cambrian Period is marked by the first evidence of organisms, such as trilobites, with well-developed eyes.

Another possibility is that ocean chemistry changed significantly and the seas became richer in dissolved calcium and phosphorus. Animal life in the sea had to find a way to rid their bodies of calcium and phosphate salts. Instead of secreting these substances directly into the environment, the animal could have deposited them around their outermost membranes. These organisms thus found themselves with hard parts that could be useful for support and protection.

Once an organism developed the ability to armor itself with a mineralized shell, this body armor would be a great advantage in the struggle for survival as predation became

FIGURE 12.11 Reconstruction of the early Cambrian sea floor of about 540 million years ago. The flowerlike structures are the tentacle-bearing mouth regions of corals. In the foreground is a large trilobite, a distant relative of modern crustaceans like crabs and lobsters. Smaller trilobites are seen swimming on the right. American Museum of Natural History

established. Whatever the causes, about 540 million years ago external skeletons became widespread among the life forms in the sea (Figure 12.11).

The variety of early animal body plans is apparent in fossils recovered from an unusual 520-million-year-old deposit called the Burgess Shale. These muddy sediments represent an undersea landslide that entombed an array of Cambrian age organisms. The fossil beds, now found high in the Canadian

FIGURE 12.12 Some of the unusual arthropod-like animals found as fossils in the Cambrian Burgess Shale: (a) *Anomalocaris*, a predator with large front appendages and a rasplike mouth; (b) *Marrella*, an animal with long antennae and spines; (c) *Opabinia*, an animal with five eyes and a single long grasping appendage; and (d) *Hallucigenia*, a multi-legged creature with long spines on its back.

© 2000 John Wiley & Sons, Inc. New York. Levin, *Earth Through Time*, 7th edition, Figure 10.9. This material is used by permission of John Wiley & Sons, Inc.

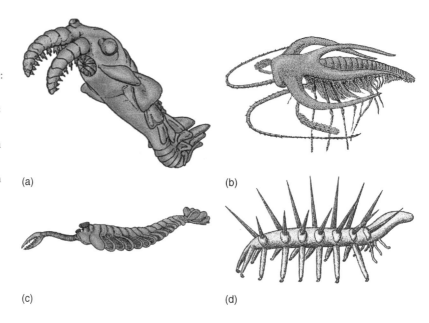

(a)

(b)

(c)

(d)

Rockies, preserve a rich diversity of animals with unusually well-preserved soft parts. Some of these creatures resemble well-known animals forms, but others are unusual (Figure 12.12) and may represent evolutionary experiments with body plans that were not viable in the long run.

12.5 The fishes

During the Cambrian explosion of life (about 540 million years ago), one group of bilaterally symmetrical wormlike animals developed the beginnings of an *internal skeleton* – initially a rod of stiffened cartilage running down the length of the organism (Figure 12.13). *Pikaia*, an animal found in the 520-million-year-old Burgess Shale may be an early example of this line of evolution. This cartilaginous rod or *notochord* later evolved into a calcified segmented backbone providing points of attachment for muscles. These first animals with internal skeletons were primitive fishes. The fishes could move through the water rapidly by thrusting their bodies from side to side in a whiplike motion.

The primitive fishes were heavily armored, and were also jawless, with only a simple opening on their underside for a mouth (Figure 12.14). The jawless fishes lived by shoveling

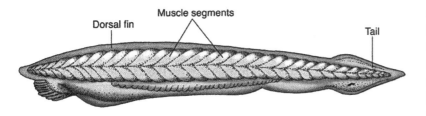

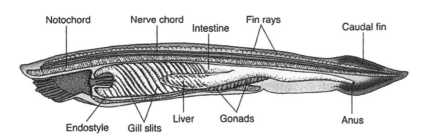

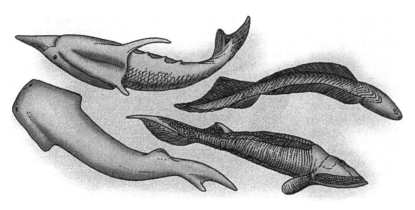

FIGURE 12.13 The lancelet (or amphioxus), a primitive modern "vertebrate" with a notochord of stiffened cartilage running down the length of the animal. The earliest fishes probably resembled the lancelet. The length is about 3 inches (7.5 centimeters).

© 2000 John Wiley & Sons, Inc. New York. Levin, *Earth Through Time*, 7th edition, Figure 10.10. This material is used by permission of John Wiley & Sons, Inc.

FIGURE 12.14 Reconstruction of jawless fishes that lived about 400 million years ago (in the Early Devonian Period). The shovel-like heads, asymmetrical tails, and the flattened bellies of these fishes probably relate to a life on the ocean bottom. These organisms averaged about 8 to 12 inches (20 to 30 centimeters) in length.

© 2000 John Wiley & Sons, Inc. New York. Levin, *Earth Through Time*, 7th edition, Figure 10.11. This material is used by permission of John Wiley & Sons, Inc.

through the muddy sea floor with their heads, ingesting mud and passing the organic-rich sediments through their guts.

By about 400 million years ago (in the Early Devonian Period), a branch of the fishes developed hinged jaws. Jaws originated through changes in the sets of arch-shaped bones that support the membranes of the gills. The development of jaws opened up a host of new opportunities for making a living, including the development of a predatory mode of life. Some of the early armored jawed fish grew to several meters in length and were the most formidable predators of their day. Other fishes with less armor used speed and agility to escape from their armored predators. Eventually the heavily armored fishes died out and the more agile fishes with more lightly built internal skeletons proliferated.

12.6 Invasion of the land

The next great steps in the history of life involved the move from the sea to the land – a momentous development that culminated in the evolution of complex creatures, and eventually humans. Simple plants related to green algae and insects began to move from the water onto the land about 450 million years ago. The forerunners of insects evolved from sea-dwelling invertebrates that already possessed jointed appendages that were well suited for the move onto the land. In some insects, wings evolved from appendages that originally functioned as gills. Plants developed the kind of stiff support, internal water transport system, and reproductive strategies required for life out of water, and rapidly grew in size over the next 20 million years.

Another factor may have influenced the development of life on land. Prior to about 450 million years ago, the ancestral continents that are now North America, Europe, and Africa were separated by an ocean. This ocean pre-dated the formation of Pangaea and the creation of the present Atlantic Ocean. Geologists call the ancient ocean the Iapetus Ocean (or the Proto-Atlantic) as it occupied virtually the same position as the present Atlantic.

The story of the Iapetus Ocean goes back about 600 million years. At that time, North America, Africa, and Europe were joined together in a super-continent. A rift or split developed in the lithospheric plate carrying these continents, and ancestral North America moved away from its continental neighbors to the east. Sea-floor spreading over the next 150 million years created a wide ocean basin – the Iapetus Ocean. Then, about 450 million years ago, subduction zones developed on one or both borders of the ocean, and the continents on the shores of the Iapetus started to approach one another again; the Iapetus Ocean began to close up.

By about 360 million years ago, the Iapetus Ocean closed, the ancestral continental plates of North America and Europe collided, and the Appalachian Mountains rose up in a region where previously only shallow seas existed. Great river systems with broad flood plains drained the mountains and large areas of coastal lowlands were formed.

The rising mountains disturbed the flow of winds and the tropical climate began to change from uniformly warm and wet to uneven, with a maximum of seasonal variation in both rainfall and temperature. In summer, the new land areas heated up more than the surrounding waters, and warm air rose over the continent pulling moist air in from the oceans. As these moist winds were forced to rise and cool in passing over the mountains, the moisture in the air condensed and summer rainfall was plentiful.

By contrast, during the winter months the land area cooled, and relatively cold, dry continental air sank and moved outward from land to sea. During this season, the land received little rainfall. The land mass was alternately wet for six months, and dry for the other six months of the year.

The collision of the two continental plates created an expanded niche for land life, but it also created the problem of seasonal inundation and drought. Fishes living in streams and ponds within the new Appalachian landmass now had the opportunity to move out onto the land, and the impetus to do so. But how could fishes make the move from life in the water to life on land?

One branch of the fishes, called *Sarcopterygians* (Figure 12.15), had stumpy, muscular fins, probably because they lived in the shallows, lying in wait, and then lunging at their prey.

Some scientists envision that as the area of wetlands decreased during periods of seasonal drought, this group of fishes could use their strong fins to move across land and seek other sources of water.[2] These fish had a second advantageous trait for life on land: they had developed lungs as well as gills so that they could gulp air at the surface during times when shallow water ponds became stagnant. In most modern fish, these lungs have since ceased to function for breathing; they have now become hydrostatic bladders used by the fishes to

---

[2] Recent fossil discoveries by paleontologist Jenny Clack of Cambridge University suggest that some transitional forms between the fishes and amphibians developed paddle-shaped limbs while still largely aquatic.

FIGURE 12.15 A lobe-finned fish (below), showing the strong muscular fins, and an early amphibian (above) with well-developed limbs.

From *Historical Geology – Evolution and Life Through Time* by Wicander, Reed and Monroe. © 2004 Thomson/ Brooks/Cole, Australia

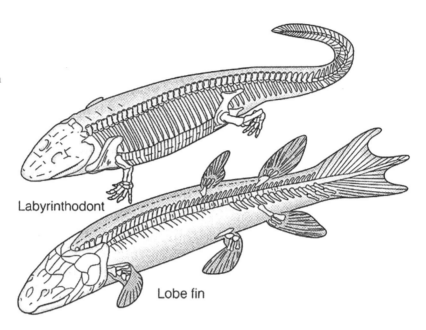

Labyrinthodont

Lobe fin

control their depth in the water. Some fish, however, still can gulp bubbles of air when their waters become oxygen deficient.

The move of the fishes onto the land did not happen overnight; it took 20 million years, and probably a like number of generations. By the end of that time the stumpy-finned fish had been transformed into an air-breathing, land-living amphibian (Figure 12.15). A comparison of the anatomy of these two creatures shows how the fins of the fish became the limbs of the amphibian. Every bone in the fin of the stumpy-finned fish is matched by a bone in the limb of the amphibian. Other changes took place in the structure of the skull, backbone, ribs, and pelvis, providing these animals with the increased support that they needed once out of the water. One of the bones that originally provided support for the gills became the *stapes*, a bone that was ultimately transformed into an important part of the terrestrial vertebrate ear.

However, the porous skin of the early amphibians could not retain moisture. These animals had to remain near water and submerge themselves from time to time, to keep from becoming dehydrated. The eggs of amphibians also lack an insulating shell for holding moisture, so these creatures were forced to lay their eggs in water.

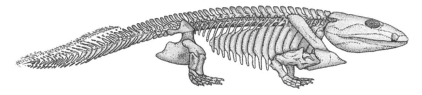

FIGURE 12.16 An early amphibian, *Ichthyostega*, lived about 360 million years ago. This animal was about 3 feet (1 meter) long. *Ichthyostega* was one of the first four-legged animals or "tetrapods".

From *Life Through Time* by H. L. Levin. © 1975 William C. Brown Co.

FIGURE 12.17 The coal swamps of the Carboniferous Period about 325 million years ago during the peak of amphibian evolution. American Museum of Natural History

The new four-legged amphibians found rich sources of food in land plants and insects. They prospered, some growing to the size of modern-day crocodiles (Figure 12.16). These animals dominated the terrestrial scene for almost 50 million years. At the same time, plant life also prospered. Some plants, resembling modern ferns, grew to the size of large trees. During this interval (from about 350 to 300 million years ago, in the Carboniferous Period), swamps became widespread on the continents, creating masses of rotting vegetation that were eventually buried and compacted to form the enormous coal beds that are the source of much of today's fossil fuels (Figure 12.17).

As plate movement continued, the continents became welded into the super-continent of Pangaea. Sea level dropped, and the growing landmass of Pangaea emerged from beneath the shallow seas. With the emergence of new land, the climate began to change. Major changes were also taking place in the evolution of the vertebrates.

12.7 Summary

By the end of this chapter you should be familiar with the following concepts and topics:

The early history of life on Earth
The development of life
   The cell
   Photosynthesis: growth of atmospheric oxygen
Development of multi-cellular organisms: evidence from
   fossil records
   The tree of life
   First cells with nuclei
   Use of oxygen
   Origin of sex
   Multi-cellular organisms
The Cambrian explosion
   Simple hard-bodied organisms
   Fishes
   Invasion of the land

Questions
12.1. When did the first hard-bodied animals appear and why?
12.2. Describe the events leading from the evolution of the first fish to the fish leaving the water.
12.3. What are the advantages of amphibian life over that of fishes?
12.4. What were the geological circumstances that led to the invasion of the land by the fishes?

CHAPTER 13

The development of higher life forms

The age of the reptiles and dinosaurs

As the vast land mass of Pangaea began to assemble, the interior of the super-continent became drier. Developments that allowed animals to move away from water courses and swampy areas to explore the new regions were favored in the evolutionary process. By about 300 million years ago, a new animal had evolved from the amphibian rootstock. This new creature, with leathery skin and watertight eggs was the first reptile. Over the next 100 million years, two major branches of reptile evolution led to the dinosaurs and to the first mammals. Evolving in step with the changing climate of Earth and developing traits to enable them to exploit an array of environments, these early land animals were amazingly complex and diverse.

Living on land requires an altogether different set of characteristics to living in water. In this chapter we will follow the evolution of reptiles from the early amphibians, and see how they evolved body characteristics suitable to a dry land environment. We will learn how the evolution of life is inextricably linked to the Earth's climate, and how life has had to constantly adapt to its changing environment. In a relatively stable climate, life forms such as the dinosaurs had the chance to diversify and flourish. We will look at the amazing variety of dinosaurs that dominated the Earth for around 135 million years, and at why they eventually and so suddenly became extinct.

13.1 Evolution of the reptile

Through a succession of mutations, traits favorable in the new land environments had begun to evolve in the time of the amphibians. In a world where access to water was becoming more difficult, a tough membrane to seal in body moisture was a trait advantageous for survival. Leathery skin to hold in moisture enhanced the ability of the possessor to venture into the new, drier continental habitats. The development was probably a gradual process. In each clutch of young amphibians the changing environment selected those best able to resist dehydration in order to survive and reproduce. Amphibian eggs, with gelatinous coverings that dried out easily had to be laid in water. Life on land required eggs that were covered with a watertight coating (Figure 13.1). By about 300 million years ago, a new animal had evolved – one that lived entirely on land, possessed a leathery skin and produced eggs resistant to dehydration. The early amphibians had evolved into the first reptiles (Figure 13.2).

Over the next 50 million years, from about 300 to 250 million years ago, as Pangaea grew in size, the reptiles proliferated. The geological changes that occurred during that interval greatly affected Earth's climate. These climatic changes would determine the destiny of the reptiles and their progeny.

As the super-continent Pangaea assembled, the rate of sea-floor spreading diminished and the level of the sea dropped.

Previous page: *Euoplocephalus* (a type of Ankylosaur) with Stygimoloch Group.
© Phil Wilson

FIGURE 13.1 Fossilized reptile egg with a leathery outer shell that allowed it to be laid out of the water.
Mark Mohell and the Australian Department of the Environment, Water, Heritage and the Arts

(a)   (b)

FIGURE 13.2 *Seymouria*, one of the earliest reptiles (about 300 million years ago), reconstructed from a nearly complete fossil skeleton found near Seymour, Texas. The animal was about 20 inches (50 centimeters) long. *Seymouria* is a link between the amphibians and the reptiles.

The climate began to cool. The development and spread of land plants may have contributed to this cooling of global climate. Land plants with deep root systems pump carbon dioxide into the soil, where it accelerates rock weathering. This process increases the rate of removal of carbon dioxide from the atmosphere, tending to cool the climate.

During this time, a large ice cap grew to cover portions of what are today the continents of Antarctica, Australia, southern Africa, southern South America, and the Indian subcontinent. These continents were joined together to form the southern or *Gondwana* portion of the super continent (Figure 13.3). As discussed earlier, the ice sheet reflected more solar energy back to space causing further cooling of the climate.

This major change in climate had an important effect on the course of evolution. Animals living on the margins of the Gondwana ice sheet evolved traits that allowed them to continue to function in cooler climates. Of primary importance is the ability to keep body temperature high (and thus maintain the

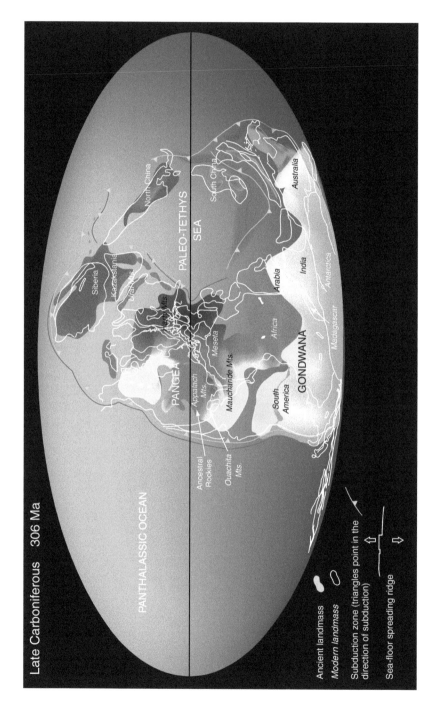

FIGURE 13.3 The configuration of the continents at the time of the Late Paleozoic Ice Age. The position of the large ice sheet on the Gondwana super-continent in the Southern Hemisphere has been inferred from glacial deposits and scratched and grooved rocks dating from this time.

Christopher R. Scotese. PALEOMAP Project, Texas

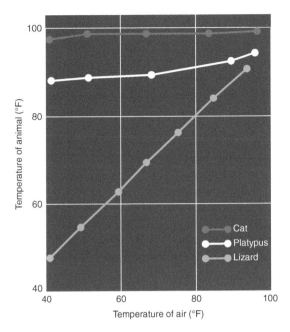

FIGURE 13.4 The meaning of warm-bloodedness. The cat is a typical warm-blooded mammal, with many special ways of keeping its body temperature constant. Sweat glands or panting cool a warm-blooded animal when the air is hot, and the blood vessels under the skin dilate to carry body heat to the surface, where it is lost. Because of these defenses, the temperature of the cat scarcely changes when the air temperature increases from 40 °F to 95 °F (5 °C to 35 °C).

The lizard – a typical reptile – is cold-blooded and lacks the cat's defenses against temperature change. Its body temperature is usually close to the temperature of the outside air. The duck-billed platypus – a link between the mammals and the reptiles – controls its body temperature more effectively than the lizard but not as well as the cat.

chemical reactions needed for cellular activity and growth) even when air temperatures dipped well below freezing. In vertebrates that have developed internal means of regulating their body temperature, this ability is called *endothermy*, as opposed to *ectothermy*, in which an animal has no internal means of regulating its body temperature. These traits are commonly called warm-bloodedness and cold-bloodedness (Figure 13.4).

Modern reptiles are ectotherms – they rely on the heat of the Sun (or a warm patch of ground) to raise their body temperature to a point at which they can become active. At night, or in cool weather, as their body temperature drops, ectothermic animals must retreat temporarily into burrows or hibernate. It seems likely that the early reptiles were also ectothermic. During the cooling that accompanied the Late Paleozoic Ice Age, however, a new kind of animal developed, one that had the beginnings of characteristics that we usually associate with endothermy.

13.2 The mammal-like reptiles

About 275 million years ago, the sail-back reptile *Dimetrodon* made its first appearance in the fossil record (Figure 13.5).

FIGURE 13.5 *Dimetrodon* – the sail-back reptile. *Dimetrodon* was distinguished by the large "sail" on its back, which seems to have been an early design for regulation of body temperature. *Dimetrodon* was about 10 feet (3 meters) long.

© 1997 M. Moravec, Dinosaur Corporation

What was the reason for such a body structure? Scientists now believe that the sail-like structure was designed to help the animal maintain a relatively constant body temperature. During the day the sail could be spread out and turned towards the Sun to collect solar energy. At other times the sail could be pulled back to conserve body heat. The sail-backs had developed a primitive temperature regulation system.

Another piece of evidence that the sail-backs were on the path toward warm-bloodedness is the nature of their teeth. The early reptiles possessed peglike teeth. The teeth of *Dimetrodon*, however, show the first signs of differentiation into cutting teeth toward the front of the jaw, "canine" saber-like teeth at the sides, and flatter, grinding teeth in the back. This is the beginning of a pattern of dentition that was to evolve into the specialized teeth of the mammals. *Dimetrodon* is one of the earliest in the line of so-called *mammal-like reptiles* that eventually gave rise to true mammals.

What was the reason for specialized teeth in the early mammal-like reptiles? The answer may have been that the mammal-like reptiles were developing a higher metabolic rate. Specialized teeth are only needed if an animal must chew its food efficiently. Modern mammals chew their food into small pieces before swallowing. The mammal must digest its food rapidly in order to extract the maximum amount of energy to stoke its internal fires. Modern reptiles, by contrast, swallow their food whole, or bite it off in large chunks. Since they are ectotherms, they have little need to digest their food rapidly. Reptiles remain inactive as they slowly digest their meal.

By about 255 million years ago (in the Late Permian Period) (Figure 13.6), more advanced mammal-like reptiles had evolved. *Cynognathus*, the "doglike" reptile, was even more mammalian (Figure 13.7), with differentiated teeth and a relatively larger brain than that of the earlier reptiles. This animal may have been one of the first to be covered in fur – reptilian scales modified for the purpose of insulation.

The end of the Permian Period (about 250 million years ago) of Earth history was marked by one of the most severe mass extinctions of life seen in the fossil record. In the oceans, more than 90% of the species of life died out. On land, many reptiles and amphibians and large numbers of plants became extinct. The event may have been caused by massive volcanism, or the impact of a large asteroid or comet. Great basaltic lava flows were erupted in Siberia at about this time. Some scientists have reported evidence for a large impact crater off the coast of Australia. The cause of the extinctions, however is still unknown.

13.3 The mammals

The first true mammals appear in the fossil record about 200 million years ago (in the Late Triassic Period). They were the size of a mouse (Figure 13.8). The small size of the early mammals is a sign of competition with a more successful group of animals. During the same period in which the mammals were evolving, another kind of animal appeared that was also descended from the ancestral stock of reptiles. The new kind of reptile was the dinosaur.

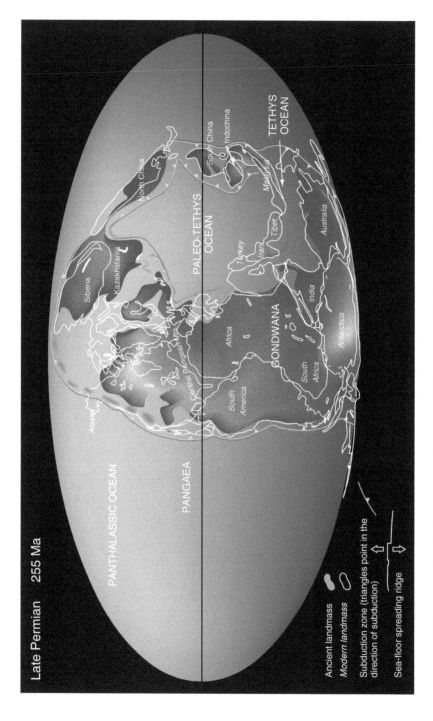

FIGURE 13.6 Reconstruction of the continents about 255 million years ago, in the Late Permian Period, showing the super-continent Pangaea.

Christopher R. Scotese, PALEOMAP Project, Texas

FIGURE 13.7 *Cynognathus* lived about 255 million years ago. Changes in the skeleton of this animal have raised its body well off the ground, although its posture still retains a hint of the sprawling gait characteristic of its reptile ancestors. *Cynognathus* was the size of a wolf. Its teeth were beginning to evolve into the canines and incisors characteristic of modern mammals, and used for biting and tearing the prey into small pieces.

FIGURE 13.8 This early mammal, known as *Morganucodon* (Morning-tooth), lived about 200 million years ago. This small animal was only about 4 inches (10 centimeters) long.
Smithsonian Institution-National Museum of Natural History, Scientific advisor Anna K. Behrensmeyer, sculpture by Gary Staab

In the branch of evolution leading to the dinosaurs, speed and agility were of the essence. These desirable traits were achieved in the dinosaur ancestors by a change in posture, in which the legs no longer sprawled outward to either side but were tucked in, with the body raised off the ground. Still another innovation appeared in the dinosaur line: The ancestral dinosaur, four-footed at the start like other reptiles, gradually evolved a two-legged posture. Their hind limbs became strong and muscular, which gave them additional speed, while their forelimbs were freed for grasping their prey.

These were the early dinosaurs. They were birdlike in appearance, but with the long tail of a lizard. A jaw with a wide gape, and many sharp teeth, completed the picture of an effective carnivore (Figure 13.9).

FIGURE 13.9 This small
reptile, about the size of a
rooster, the ancestor of the
dinosaurs, lived about 225
million years ago.

By permission of Cecilia
Barnbaum and Edward E.
Chatelain, Valdosta State
University

How did it come to pass that the mammals were surpassed by the early dinosaurs? One possibility involves changes in global climate. The line of evolution leading to the mammals originated during a time when Earth's climate was relatively cool. The mammal-like reptiles continued to thrive until inexorably Earth's climate began to change again.

Major shifts occurred at this time in the movements of Earth's plates. Not long after the final assembly of Pangaea, the super-continent began to break apart again. At that time, Pangaea was split by new rift valleys. Volcanic activity was widespread, as new ocean areas formed by sea-floor spreading. The hot, newly minted ocean ridges expanded and drove some of the ocean water onto the continents to create new shallow seas (Figure 13.10). The increased volcanic activity poured carbon dioxide into the atmosphere, and the climate grew warmer.

In the new warmer climate, cold-blooded animals, like the early dinosaurs, could remain active. Warm-bloodedness, once a valuable trait, became a liability since there was less of a need to keep warm, and endotherms must eat often to maintain their high body temperatures. Competition with the dinosaurs led to a trend toward smaller size in the mammal-like reptiles and their mammal descendants. The small mammals were able to hide from their dinosaur predators and scavenge enough food to survive. They were fated to spend the next 135 million years as subordinates in a world ruled by the giant dinosaurs.

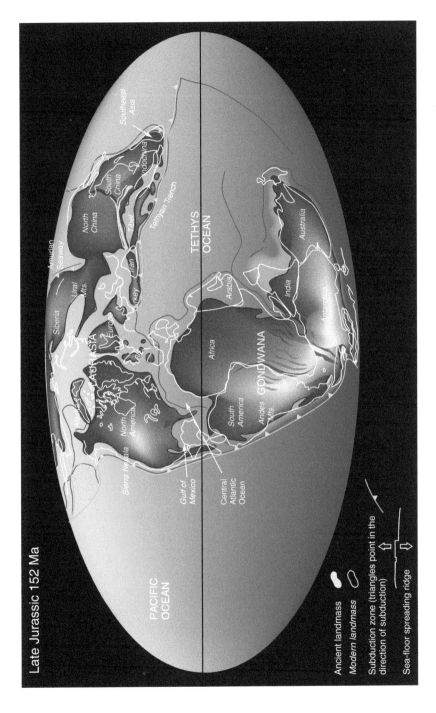

FIGURE 13.10 Continental reconstruction for the Late Jurassic Period, about 152 million years ago. The presence of inland seas and increased atmospheric carbon dioxide from renewed volcanism warmed the Jurassic climate.

Christopher R. Scotese, PALEOMAP Project, Texas

Another factor in the replacement of the mammal-like reptiles by the dinosaurs was a severe mass extinction of life that occurred about 200 million years ago, at the end of the Triassic Period. This event was marked by the disappearance of many forms of life in the oceans, and also led to extinctions of many land-dwelling reptiles. Many lines of reptiles died out; others survived, including the early dinosaurs.

13.4 The dinosaurs

For more than 100 million years Earth enjoyed a climate of unparalleled warmth, and the dinosaurs and other reptiles proliferated. They flowered into a variety of creatures that ruled over the air, the sea, and the land. Flying reptiles soared and glided, ready to swoop down on smaller prey; some had wing-spans as great as 50 feet (17 meters) – the size of a small aircraft (Figure 13.11). Giant crocodiles with jaws 6 feet (2 meters) long lurked in the waters; turtles weighing 3 tons paddled across the inland seas of North America (Figure 13.12).

The greatest giants lived on the land. These were the dinosaurs. The earliest dinosaurs were small carnivores, but they rapidly evolved into new forms and many grew larger. By about 200 million years ago, all of the major groups of dinosaurs were in existence. Some of the early bipedal carnivores developed into large meat eaters such as *Allosaurus* by about 150 million years ago (in the Jurassic Period) and eventually into the largest meat-eating dinosaur of all, *Tyrannosaurus rex*, about 70 million years ago (in the Late Cretaceous Period). *Tyrannosaurus rex* grew to a length of 50 feet (17 meters) and a weight of 10 tons, with jaws filled with daggery-like teeth (Figure 13.13).

Among the largest dinosaurs were huge herbivorous dinosaurs like *Apatosaurus* (Figure 13.14). These large plant-eating dinosaurs dropped back onto all four legs and, like the modern-day elephant, were protected by their bulk from most predators. Plant-eaters were also represented by the *Ornithischian* dinosaurs, such as *Stegosaurus* (Figure 13.15), and the armored *Ankylosaurs*. Later, horned dinosaurs such as *Triceratops* (Figure 13.16) and the duck-billed dinosaurs (Figure 13.17) became common.

FIGURE 13.11 During the age of dinosaurs, the flying reptiles flourished. The largest flying reptiles soared on wings with a span of 30 to 50 feet (10 to 17 meters).

Joe Tucciarone and Jeff Polling

FIGURE 13.12 During the time of the dinosaurs, there were also giant reptiles in the sea. *Mosasaurs* were giant marine reptiles 40 feet (13 meters) long.

Daniel W. Varner

FIGURE 13.13 The meat-eating dinosaur *Tyrannosaurus rex*, with a length of about 50 feet (17 meters), lived during the Late Cretaceous Period (about 70 million years ago). Here, a group of Tyrannosaurs threatens the horned dinosaur *Styracosaurus*.
   Phil Wilson

FIGURE 13.14 The largest dinosaurs were plant-eaters, such as 65-foot (20-meter) long *Apatosaurus*. *Apatosaurus* lived in western North America, then a tropical land of forests and swamps, during the Late Jurassic Period about 150 million years ago.
   Phil Wilson

One group of the small carnivorous dinosaurs gave rise to the birds. The earliest well-known fossil bird, *Archaeopteryx*, which appeared about 150 million years ago (in the Jurassic Period), had the skeleton of a small dinosaur, including a toothed skull, bony tail, and claws on its wings, but was

FIGURE 13.15 *Stegosaurus*, about 20 feet (7 meters) long, lived in the Late Jurassic Period about 150 million years ago. The plates at the back of this creature may have acted as cooling vents to help maintain a stable body temperature. The large tail spikes clearly were for defense. *Stegosaurus* had a brain the size of a walnut.
Phil Wilson

FIGURE 13.16 *Triceratops*, one of the largest horned dinosaurs, was 15 feet (5 meters) long. *Triceratops* lived during the Late Cretaceous Period about 70 million years ago.
Phil Wilson

covered by true feathers (Figure 13.18). A number of exceptionally well-preserved small, feathered dinosaurs have been found in China, showing that feathers were not an uncommon trait among dinosaurs. Feathers probably evolved for insulation, keeping the small dinosaurs warm during the cool evenings.

FIGURE 13.17 Duck-billed dinosaurs, like this *Hadrosaurus*, were abundant during the Late Cretaceous Period, about 70 million years ago.

Phil Wilson

FIGURE 13.18 *Archaeopteryx*. In the line of reptilian evolution that led to the birds, the edges of the reptile's horny scales became more and more ragged, and gradually the scales changed into feathers. Feathers grew from the forearm to make a wing; they also sprouted from the long, lizardlike tail of the animal. This creature, *Archaeopteryx*, half reptile and half bird, first appeared in the fossil record about 150 million years ago (in the Jurassic Period) and was the ancestor of the modern bird.

Karen Carr

13.5 Why were the dinosaurs successful?

The dinosaurs were among the most successful animals ever to walk Earth. For approximately 135 million years these creatures, which ranged from chicken-sized reptiles to the largest land animals of all time, dominated Earth. During this long interval, Earth's climate was particularly warm and equable. Continental areas were free of ice sheets, and most of the world had a tropical or subtropical climate.

For most of this period, large areas of the continents were submerged beneath warm shallow seas. Large amounts of carbon dioxide were added to the atmosphere by increased volcanic activity at mid-ocean ridges and subduction zones.

This created a world with perhaps ten times more carbon dioxide than in the present atmosphere, with a more intense greenhouse effect and warmer climate.

Another factor that led to the warm climatic conditions during the Mesozoic Era was the low relief of the terrain, and the general lack of high mountains. The Appalachians had been worn down by erosion, and the great mountain belts of the modern world – the Himalayas, Rockies, Alps, and Andes – were either in the early stages of their development, or had not yet begun to form. The lack of mountain barriers allowed for the freer circulation of air currents between the equatorial and polar regions. The transport of warm air and warm ocean currents toward the polar regions warmed the climate in these areas, and made for a more even temperature over the planet.

Under such conditions of almost universal warmth and abundant rainfall, lush tropical and subtropical forests and marshlands came to cover large areas of the exposed continents. This environment created conditions in which the dinosaurs flourished. Many of the dinosaurs show evidence of complex mating behavior, parental care, and a social structure in which they lived in large herds and nesting colonies. Some dinosaurs were apparently able to stabilize their body temperatures, either as a result of their large size or relatively high metabolism. But elaborate mammalian types of warm-bloodedness and high intelligence were not required for success in the Mesozoic climate regime.

Beginning about 80 million years ago, however, the environment began to change. The rate of sea-floor spreading began to slow, the ocean ridges partly subsided, and the waters of the shallow seas began to drain back into the deepening ocean basins. The exposure of more land area caused the climate to become cooler and more seasonal. This is because the land absorbs heat in the summer, but loses it rapidly in winter so that the seasons tend to be extreme with cold winters. As weathering rates of the newly exposed continental rocks increased, the carbon dioxide levels in the atmosphere began to fall. Meanwhile, plate movements continued, progressively carrying the continents into higher latitudes and therefore leading to further cooling of the dinosaurs' habitats.

The result of all these factors was the beginning of a global cooling and an increase in seasonality of climate. Some pale-ontologists believe that the dinosaurs were already too specialized – as a result of 100 million years of rather stable climate conditions – to cope with these ongoing changes in the environment. But the dinosaurs disappeared abruptly 65 million years ago in a dramatic event that was unrelated to the long-term environmental changes (Figure 13.19). More than 75% of the species of life on Earth became extinct at that time.

13.6 The End Cretaceous extinction

Sixty-five million years ago, many species of life in the seas, from one-celled algae to giant marine reptiles, became extinct. On land, the dinosaurs vanished, and forest plants were deci-mated. In the air, the flying reptiles disappeared.

Geologists recognize this mass extinction as marking the end of the Mesozoic Era. Rocks below and older than the boundary belong to the Cretaceous Period, the final chapter in the "Age of Reptiles," and rocks above, and therefore younger than the boundary, are assigned to the Tertiary Period of Earth history, when birds and mammals became dominant.

The cause of this wave of extinction has long been one of the outstanding problems for students of the history of life. Some scientists suggest that the extinctions were abrupt. An abrupt extinction implies some sort of global catastrophe. Other scientists hold that the extinctions happened gradually over millions of years. Such a gradual extinction implies progressive environmental changes over time as the cause.

13.7 The Alvarez Hypothesis

One of the best places to study the Cretaceous extinction is in rock layers exposed in the walls of a deep gorge near the medieval Italian town of Gubbio. These layers of limestone are composed of countless shells of tiny floating marine organisms, or plankton, that were deposited on the ancient

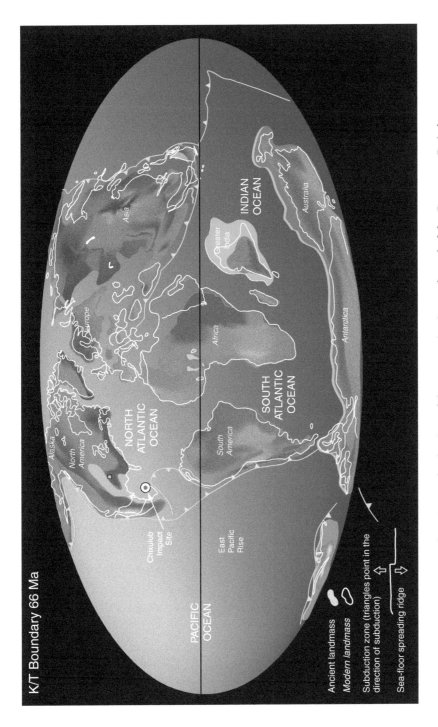

FIGURE 13.19 Reconstruction of continents, for the time of the mass extinction at the end of the Cretaceous Period, 65 million years ago. The continents are still partly flooded with shallow seas. The site of the asteroid impact on the Yucatán Peninsula is shown by the small circle.

Christopher R. Scotese, PALEOMAP Project, Texas

FIGURE 13.20 Photograph of the boundary between the Cretaceous and Tertiary Periods (65 million years ago) in a gorge near the Italian town of Gubbio. Cretaceous limestones are overlain by darker Tertiary limestones. The boundary is marked by a thin layer of clay at the head of the hammer.

M. R. Rampino

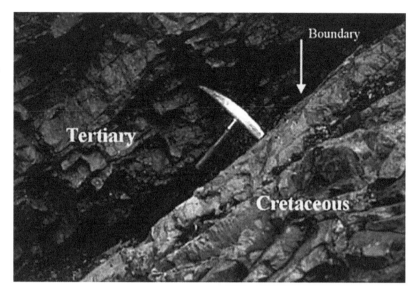

sea floor. A drastic change in the fossils in these rock layers occurs at a bed marked by a one-centimeter-thick layer of soft clay (Figure 13.20).

In 1980, a group of scientists from the University of California at Berkeley led by the late Nobel laureate physicist Luis Alvarez and his son Walter, a well-known geologist, discovered that the thin clay layer that was laid down at the time of the extinction event contained high levels of the rare element iridium, a member of the platinum group of elements.

Iridium is a rare element in the rocks of Earth's crust, because iridium has a strong affinity for metallic iron, and most of Earth's complement of iridium sank into the core during the early melting and differentiation of our planet. Most meteorites, and the asteroids and comets from which they come, are almost unmodified pieces of planetary matter, and thus contain their original abundance of iridium.

The Alvarez group's analysis showed that the source for the iridium layer was fallout of the dust produced by the collision of a 6-mile (10-kilometer) diameter asteroid or comet. Such an impact would create a large crater spewing out a global dust cloud of pulverized rock and vaporized asteroid. The mass extinction of life was caused by the dense impact-induced global dust cloud that shut out sunlight, curtailed photosynthesis, and cooled the climate. The same iridium-rich layer has

FIGURE 13.21 Photograph of the 65-million-year-old impact layer in Colorado, in sediments deposited on a river floodplain. The clay layer occurs at the base of a thin bed of coal. Dinosaur footprints are found just below the clay layer, but no dinosaur fossils have ever been found above the layer.

  M. R. Rampino

since been found at more than a hundred sites around the world (Figure 13.21).

Many scientists were initially skeptical of these findings. Some suggested that the iridium layer came from a large volcanic eruption, and they showed that some volcanoes, such as those on the island of Hawaii, release iridium brought up from deep within Earth. Massive lava flow eruptions were taking place in India just at the end of the Cretaceous Period, and some scientists still think that these eruptions caused climatic cooling that led to the extinction.

The thin clay layer, however, also contains tiny grains of the mineral quartz with unusual sets of crisscrossing planes (Figure 13.22). Studies show that these quartz grains must have been subjected to the extremely high pressures generated by a large-body impact. The layer has also yielded glassy beads that originated as droplets of molten rock thrown out of the large impact crater.

If a large impact occurred 65 million years ago, then a huge impact crater should exist somewhere on Earth. Calculations and experiments suggested that the crater would be about 20 times the diameter of the impacting object, or about 120 miles (200 kilometers) in diameter. In the 1980s, exploration for deposits of oil and natural gas in Mexico turned up evidence

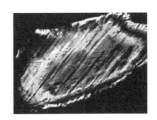

FIGURE 13.22 A sand-sized grain of quartz taken from the clay layer that occurs at the time of the mass extinction, 65 million years ago. The crossing lines are planes in the mineral grain filled with amorphous quartz of extremely high pressure origin.

  Richard P. Walker, Shawn T. Smith, and Steven M. Smith, USGS

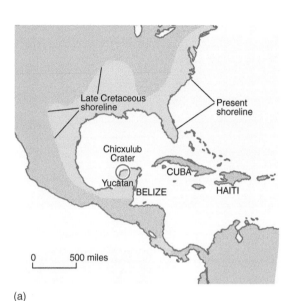

(a)

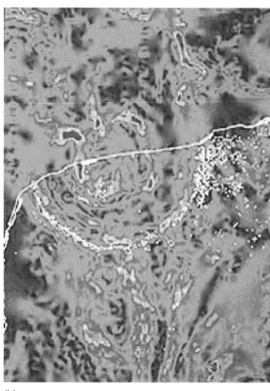

(b)

FIGURE 13.23 (a) Location of the large (∼100 miles or 160 kilometers in diameter) Chicxulub impact crater (circle) in the Yucatán Peninsula of Mexico. Figure 13.23(b) The Chicxulub crater is imaged by using measurements of the strength of Earth's gravity over the buried crater. The white line is the coastline of the Yucatán Peninsula. Half of the crater is offshore. The white dots are small lakes, or *cenotes*, that occupy sinkholes created by dissolution of limestone along the circular fractures related to the crater.

Alan Hildebrand, University of Calgary

for a large, buried multi-ringed crater underneath the Yucatán Peninsula (Figure 13.23). Drilling at the site revealed melted and shocked debris that formed 65 million years ago. This crater, named Chicxulub after a small town near its center, is the source of the worldwide layer of debris that occurs at the time of the mass extinction.

13.8 Mass extinctions in Earth history

The Cretaceous event was one of a number of mass extinctions that punctuate the fossil record of life. The record of extinctions of marine organisms for the past 540 million years shows at least five major mass extinction events (Figure 13.24). Traditionally, these extinction episodes have been attributed to terrestrial causes such as changes in global climate or volcanism. The recognition that the End Cretaceous extinction was caused by an impact leads to the

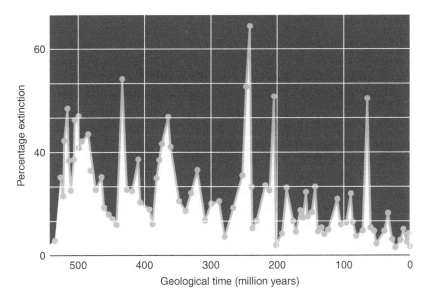

FIGURE 13.24 The record of mass extinctions. The percentage of marine organisms that became extinct at various times during the last 540 million years. The peaks represent extinction events of various magnitudes. The most severe mass extinction occurred at the end of the Permian Period, 250 million years ago. The large extinction peak at 65 million years ago marks the Cretaceous–Tertiary boundary.

question: Could there be a general connection between mass extinctions and impacts?

Astronomical observations of Earth-crossing asteroids and comets, and the records of the inner planets predict that Earth will be hit by a 6-mile (10-kilometer) diameter asteroid or comet – the estimated size of the impactor at the end of the Cretaceous – about once every 100 million years. These observations can be used to predict a record of impact-induced mass extinctions during the past 540 million years. The expected impact history for this time period predicts about five major mass extinctions. This agrees well with the record of extinctions. The result suggests that the record of extinction events could be explained by comet and asteroid impacts. This hypothesis is testable, in principle, by searching for evidence of impacts at extinction episodes.

The discovery of the impact layer at the time of the mass extinction at the end of the Cretaceous prompted the search for impact signatures at other geological boundaries. A widespread iridium anomaly was found at the time of a lesser extinction that occurred about 36 million years ago near the end of the Eocene Epoch of Earth history. Associated with this iridium peak, researchers discovered shocked quartz grains and tiny glass objects called *microtektites*, originally droplets formed when an impact blasts the surface rocks into a molten spray.

Geologists at the US Geological Survey recently found a 50-mile (90-kilometer) diameter crater beneath Chesapeake Bay in the eastern United States that dates from 36 million years ago. A similar large crater of the same age, called Popigaí, was discovered in Siberia. The search for evidence of impact at the times of other mass extinctions is a focus of current research.

13.9 Catastrophism

The outcome of this new line of research may be a concept of geological and biological change that relates to Earth's place within the Solar System. At the beginning of the nineteenth century, the doctrine of *catastrophism* was the vogue in geology. According to the catastrophist doctrine, the geological and fossil records were punctuated by sudden breaks, during which great paroxysms of extinction took place. Collisions of comets with Earth were suggested as a possible cause of these traumatic episodes.

James Hutton (1726–97), the Scottish scientist often called the "father of geology," is said to have discovered "deep time" – the realization that geological history represented an almost unimaginable span of time. This concept was utilized by British geologist Charles Lyell (1797–1875) in his influential *Principles of Geology* (published from 1831 to 1833) to propose what became known as the *principle of uniformitarianism*. Lyell argued that, given the great expanse of geological time, slow geological processes were sufficient to explain the major features of the geological record, such as great mountain ranges, volcanic plateaus, deep canyons, and the episodic inundation of the continents by the sea. Catastrophism was rejected as unnecessary and contrary to the geological facts.

According to Lyell, sudden breaks in the geological record should be interpreted as gaps in an incomplete and imperfect rock record, and not as cataclysms in Earth history. The abrupt mass extinctions that seemed to take place at such geological discontinuities were seen as illusions caused by the incompleteness of the rock record. As Lyell wrote, "There must be a perpetual dying out of animals and plants, not suddenly and by whole groups at once, but one after another."

Lyell went further in proposing that the rates of geological processes were slow and relatively constant throughout geological time. There was no room in Lyell's gradualism for catastrophic change.

Geological processes, like volcanic eruptions and earthquakes, vary not only in intensity but also in frequency. Small events are common, whereas larger events are rare. If we consider impacts to be another normal geological process, then we can see that they follow the same pattern. The largest impacts, which have caused mass extinctions, have left their imprint on the history of life and the course of evolution on Earth.

Mass extinctions may be necessary for evolutionary change. In the fossil record, extinctions are commonly followed by periods of rapid evolution and diversification of surviving groups of plants and animals as they evolve to fill environmental niches left open by extinct animals and plants, eventually leading to a much altered biosphere in which new species predominate. Thus, at the end of the Cretaceous, a world dominated by dinosaurs and flying reptiles was transformed into the modern world of mammals and birds.

13.10 Summary

By the end of this chapter you should be familiar with the following concepts and topics:

The development of higher forms of life
The evolution of reptiles
The mammal-like reptiles
The mammals
The dinosaurs
Evolution of life and Earth's climate
Adaptation of life to its changing environment
The extinction of the dinosaurs – Alvarez Hypothesis
Catastrophism

Questions

13.1. What was the climate like during the Mesozoic Era
(250 to 65 million years ago), and how did this affect the
evolution of the vertebrates?

13.2. Describe the circumstances that led to the extinction of
the dinosaurs.

13.3. What were mammals like during the era of the
dinosaurs?

13.4. What is "catastrophism" in geology?

PART V

Evolution of intelligent life

CHAPTER 14

The mammals

The rise of intelligence

Mammals, our own class of animal, first appeared about 200 million years ago, about the same time as the first dinosaurs. During the long Mesozoic Era that followed, they remained subordinate to the dinosaurs – small animals that hid during the day when the dinosaurs were active, and foraged in the cooler nights when the dinosaurs were torpid and their movements slow. With the extinction of the dinosaurs 65 million years ago, and nothing to fear but their own kind, the mammals spread quickly across the continents and occupied all the living spaces that the dinosaurs had left vacant. The Earth was theirs to exploit, and thus began the evolution of intelligence as the mammals fine-tuned themselves to their environments.

Following the disappearance of the dinosaurs, the fossil record shows that a small ratlike mammal evolved over a few tens of millions of years into the ancestors of all modern mammals. In this chapter we will discover the advanced characteristics that made mammals so successful in a cool climate, and why the offspring of mammals had a much greater chance of survival than their reptilian forebears. We will learn why intelligence developed in mammals to a much greater extent than in the dinosaurs with which they originally co-inhabited the planet. We then follow the evolution of mammals to apes, as they developed ever greater levels of intelligence.

14.1 Evolutionary radiation of the mammals

The mammals that coexisted with the dinosaurs were mostly small, furry creatures (Figure 14.1). With the dinosaurs' disappearance, the mammals spread quickly across the continents and occupied all the living spaces that the dinosaurs had left vacant. Some stayed on the ground; others went up into the

FIGURE 14.1 A rat-sized Mesozoic mammal, contemporary of the dinosaurs.

© 2000 John Wiley & Sons, Inc. New York. Levin, *Earth Through Time*, 7th edition, Figure 10.9. This material is used by permission of John Wiley & Sons, Inc.

Previous page: Two modern mammals: Koko, a lowland gorilla (born 1971) was taught in American Sign Language, and a kitten. © Ron Cohn/The Gorilla Foundation/koko.org

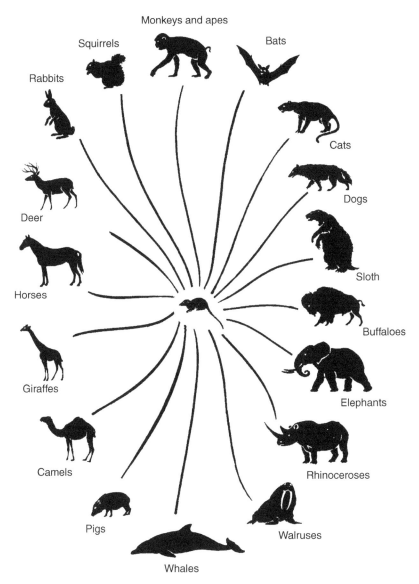

Rabbits

Squirrels

Monkeys and apes

Bats

Cats

Dogs

Sloth

Buffaloes

Deer

Horses

Giraffes

Elephants

Rhinoceroses

Camels

Pigs

Walruses

Whales

FIGURE 14.2 Evolutionary radiation of the mammals. The fossil record reveals that a small, ratlike animal – the basic placental mammal – evolved over a few tens of millions of years into the ancestors of the horse, pig, bat, whale, elephant, and other modern mammals following the disappearance of the dinosaurs. The ancestral mammals moved into all the niches – land, sea, and air – previously occupied by the dinosaurs and other reptiles.

trees; still others returned to the water. In a relatively short time, through chance mutation and natural selection, the basic mammalian stock evolved into an amazing assortment of creatures – forebears of the elephant, monkey, bat, whale, and other animals that populate Earth at the present time (Figure 14.2). This remarkable phenomenon, which has occurred many times in the history of life, is called *adaptive radiation* because many new species radiate from a common ancestor.

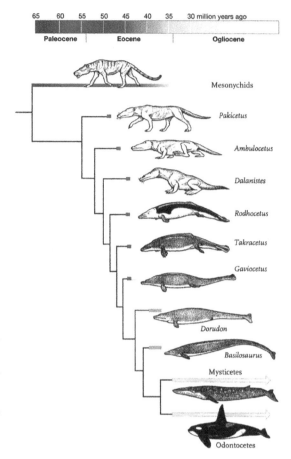

FIGURE 14.3 The evolutionary tree of whales shows the transformation of a land-living carnivore that lived about 55 million years ago into animals well adapted for life in the sea.

From p. 138 of *Evolution: The Triumph of an Idea* by Carl Zimmer. © 2001 HarperCollins Publisher, Inc. & The Doe Coover Agency

In many lines of mammalian evolution, the intermediate steps are well preserved in the geological record. In the branch of land-dwelling mammals that ultimately became whales, for example, the fossil record has revealed a progression of animals showing the transformation of a four-legged land-dwelling mammal into a semi-aquatic creature, but still retaining vestiges of rear legs. The first true whales appear in the fossil record about 35 million years ago (Figure 14.3).

The development of a branch of land-dwelling, four-legged mammals that evolved into whales is an example of *convergent evolution*. Natural selection molds the forms of life into the shapes best adapted for their environment. Animals that occupy a similar niche commonly end up with similar forms. The streamlined bodies of fish, whales, and the swimming reptiles of the Mesozoic Era are examples of convergent evolution, as are the wings of birds, bats, and flying reptiles.

14.2 Success of the mammals

The mammals were well adapted to the cooling world of the Cenozoic Era. One reason may be that mammals are warm-blooded, and better able to survive in a cool climate. When the temperature of the air is low, the fine network of blood vessels under the skin of the mammal contracts, reducing the blood to the surface and cutting down the loss of body heat. When the temperature is too high, the blood vessels expand, increasing the loss of heat from the skin and cooling the body. Many other traits contribute to warm-bloodedness; for example, an insulating coat of fur replaces the naked skin of the reptile and keeps a mammal warm when the air is cold, and the involuntary act of shivering also warms it, while sweat glands cool the body by evaporation when the temperature is high.

How did the mammals come to acquire these advantageous traits? As we have seen, the mammal-like reptiles exhibited traits that suggest the evolution of internal temperature regulation. They passed these traits on to the early mammals. The first true mammals were small animals that began to explore the possibilities of a nocturnal life. The night offered the chance of finding food and safety while the dinosaurs were less active.

Mammals possess other characteristics, in addition to warm-bloodedness, which distinguish them from their reptilian forebears. One of the most important among these is a very effective means, unique to mammals, of caring for their young. Reptiles lay their eggs and commonly display little interest in the fate of their progeny; but the mammalian mother protects the developing embryo against the hostile forces in the environment by nourishing it inside her body with her own blood; after birth, she feeds her young with milk secreted by the glands which have given mammals their name; and she continues to care for the young a long time thereafter, until they are able to fend for themselves. In these traits, one sees the beginnings of development of affection and love in the behavior of the mammals. Mammals make more effective provisions for the survival of their young than any reptile, thereby securing a great advantage in the competition for the propagation of their species.

Mammals have still other advantageous traits. For example, the jaw of a typical mammal has a set of grinding teeth, or molars, at the side, for chewing and cutting food down to a smaller size. No true reptile has specialized teeth like this. Quick replenishment of energy and a high level of continuous activity are possible with such teeth, in contrast to the inactivity of the reptile that has swallowed its prey whole. Mammals also developed a four-chambered heart, which allows more efficient oxygenation of the blood.

But the most valuable characteristic of the mammal is increased intelligence. Intelligence is the ability to plan and remember; the ability to think in the abstract as well as in concrete terms. It involves a certain amount of self-awareness and the ability to affect the environment by choice. The fossil record shows that these traits developed in the primitive mammals very early; almost from the start, they were brainier and more flexible in their behavior than the dinosaurs. When the world began to change, their traits of flexible behavior, memory, and planning became highly advantageous. This circumstance may explain the fact that, in proportion to body weight, the brains of the early mammals were three times to ten times larger than the brains of the dinosaurs. The actual size of the brain of the dinosaur was greater, of course, but nearly all this brain was needed to control its enormous body. The mammal was much smaller, and a correspondingly smaller part of its brain was needed for body control (Figure 14.4). The pint-sized mammal was the intellectual giant of its time.

14.3 The evolution of mammal intelligence

Why were the ancestral mammals brainier than the dinosaurs? This seems a puzzle, because the mammals and the

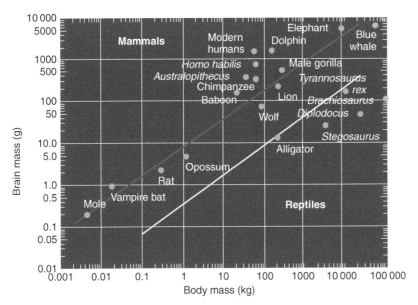

FIGURE 14.4 Chart of brain mass versus body mass for various vertebrates, living and extinct. The early mammals may be compared with the modern mole in brain size and body mass. The mammals in general plot along a line that is on average about 10 times higher in the ratio of brain to body mass than reptiles, including the dinosaurs.

dinosaurs both arose out of similar early reptilian stocks. The answer is most likely connected with the nocturnal lifestyle of the mammal, and with the fact that it used the senses of smell and hearing for survival, rather than the sense of sight.

For the dinosaur, active during the day, sight was the most important sense, and the response to a visual cue was immediate. The scene taken in by the dinosaur's eye told its brain nearly everything the brain had to know immediately, and without the need for reflection and analysis.

The brain was small in the early dinosaurs, and it remained rather small throughout their 135-million-year history. Apparently, a relatively small brain, with mostly automatic responses to visual stimuli, sufficed for the dinosaur's survival. But the mammal, forced into a nocturnal lifestyle by competition with the reptiles, could not be guided by sight during its nightly forays; it perceived the world around it through smells and sounds.

Smells and sounds are very different from visual images. A smell, for example, does not depict the object itself; it only gives a hint or trace of its presence. Perhaps the prey was here some time ago and then departed, leaving a trail of scent; it must be tracked with patience and a skill born of experience. Or the scent may have been left by a reptilian predator; then the mammal must remember the reptile's habits, and plan the night's activities accordingly.

Memory, planning, and a wisdom born of experience are critical for survival in the world of smells. The reflexive reactions of the dinosaur will not do; such immediate and direct responses to olfactory cues are rarely useful. A small brain has room only for simple circuits, commanding automatic responses. Large brains, with space available for thought, for the analysis of subtle clues, for storing the memory of past experiences, and for planning future actions, are essential to the animal that relies on smells.

Life in a world of smells places extra demands on the size of the brain in still another way. A visual image – a glance at a scene in the forest – brings a wealth of information directly to the eye. Innumerable details, in all shades of light and dark, are imprinted on the retina; everything is there, available for immediate action. But a smell contains no details; a smell is just a single thing – one particular mixture of molecules that strikes the chemical receptors in the nose. A few molecules of one kind can mean a tasty grub; a few molecules of another kind can signify a mate nearby; and a third kind may recall an entire region in the forest, some hunting ground familiar from past forays. Every detail of that forest region would be conjured up by this one odor. We humans still have this capacity, inherited from our mammal ancestors who rooted about in the dark of the night 100 million years ago. A smell can bring to mind an emotion, or a person not seen for years.

Where is the wealth of information stored, that a smell conjures up? Not in the molecule that enters the nose; a molecule does not have a picture engraved on its surface of a bog, or a stump, or a lair. The picture is in the animal's memory, waiting to be brought to life when the nose signals a familiar odor; then it springs into the conscious mind. An animal dependent on the interpretation of smells for survival must have a large brain, with a capacious memory in which the experiences of an entire lifetime are stored, like a book ready to be opened to any page on command.

A keen sense of hearing was also important to animals that moved about in the dark of night. The brain of a nocturnal

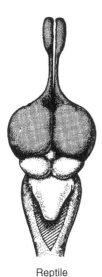
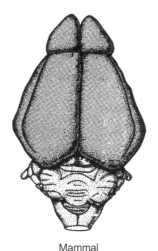

Reptile   Mammal

FIGURE 14.5 Early brain growth. Comparison of the relative sizes of the regions devoted to smell (shaded area) in the brains of a reptile (left) and a small, insect-eating mammal (right) similar to the mammals that lived in the time of the dinosaurs. The mammal's greatly enlarged smell brain is the result of the nocturnal lifestyle of this animal.

animal must have additional circuits for interpreting sounds, assessing their direction, and comparing them with information provided by other senses. This circumstance expanded the centers of the brain connected with hearing, and contributed further to the growth of the mammal's brain (Figure 14.5).

Now we can see why the brains of the early mammals became superior to the brains of the dinosaurs. Because the mammals were active at night, they were forced to memorize a map of their surroundings, with odors and sounds as their guide to the map. They could not act impetuously; they had to plan their actions. Thus, in each generation the mammals with larger brains, and greater powers of memory and planning, were more likely to survive the perils of the time and leave offspring. Throughout the years in which the dinosaurs reigned, natural selection worked steadily to prune the stock of the mammals, reducing the numbers of the small-brained and less intelligent, and augmenting the numbers of the large-brained. Over the course of time, the average brain size in the early mammals gradually increased (Figure 14.6).

Large brains, improved body design, warm-blooded metabolism, and parental care – these were advantageous traits for the early mammals. (Warm-bloodedness and brain size are

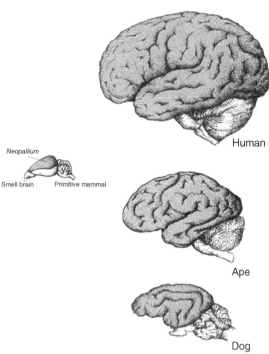

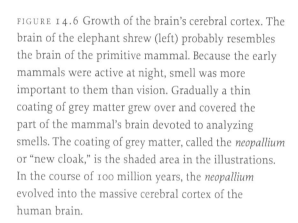

FIGURE 14.6 Growth of the brain's cerebral cortex. The brain of the elephant shrew (left) probably resembles the brain of the primitive mammal. Because the early mammals were active at night, smell was more important to them than vision. Gradually a thin coating of grey matter grew over and covered the part of the mammal's brain devoted to analyzing smells. The coating of grey matter, called the *neopallium* or "new cloak," is the shaded area in the illustrations. In the course of 100 million years, the *neopallium* evolved into the massive cerebral cortex of the human brain.

connected. Brains use up a great deal of chemical and electrical energy. An animal cannot have a large brain unless it also has a warm-blooded metabolism, to provide the brain with an abundant and continuous supply of energy.)

14.4 The tree dwellers

Among the generally intelligent mammals, one group became even more intelligent than the rest. This group lived in the trees. They had climbed up from the forest floor at least 60 million years ago, seeking food as well as protection from their enemies. The early tree dwellers or *prosimians* first appear in the fossil record about 55 million years ago. They were small animals with pointed faces and long tails (Figure 14.7).

At first, the mammals that climbed up into the trees were similar in form to their cousins who had remained on the floor of the forest, but nature set to work quickly to shape their bodies into a form better suited for survival in the new

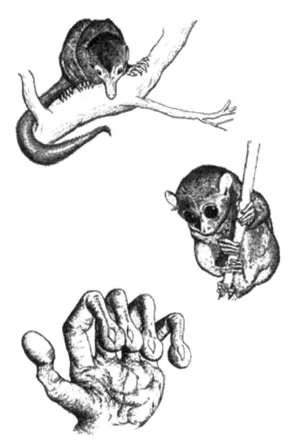

FIGURE 14.7 The tree dwellers. In the early tree dwellers (top), the eyes have started to move to the front of the head for distance judgement. The long nose of this animal betrays its recent life in the world of smells on the forest floor. The front paws still have claws. In the later stage in the evolution of the tree dweller, represented by the tarsier (middle), the eyes face full forward, providing better stereoscopic vision and distance judgement. The paw has taken the shape of a grasping hand.

Illustration by J. B. Mckoy III. Top and bottom adapted from drawings by Anthony Ravielli (From Fins to Hands).

environment of the treetops. Gradually, the pressures of that environment weeded out the animals that were least fitted for life in the trees and increased the number of those that were best fitted.

The traits needed for survival in the trees were different from those needed on the forest floor. In the new life, the greatest danger came not from other animals – for the high branches offered a safe haven from most predators – but from the risk of falling to the ground. Two physical attributes were required to reduce that risk. First, the tree-dwelling mammal needed a hand, rather than a paw, with fingers that could be curled into a tight grip around branches; second, it needed stereoscopic vision, to judge distances in leaping from branch to branch.

The tree-dwelling mammals who lacked these attributes were likely to fall to their death and leave no survivors; the ones who possessed them were more likely to live and produce offspring. Through successive generations the desirable traits of a well-developed hand and keen, stereoscopic vision, passed on from parents to progeny, came to be possessed by an ever larger part of the population, and the traits themselves were refined and strengthened. In the course of time, the paw of the original tree-dwelling animal was transformed into a hand, with supple fingers and an opposable thumb; while the eyes, originally set at the sides of the head, moved around to the front to create the overlapping field of view needed for stereoscopic vision.

In the trees, a good sense of smell was no longer useful, but keen eyesight and the ability to see vivid colors were valuable aids to the tree dweller. Among the modern mammals, only the descendants of the original tree dwellers have developed this ability to a high degree. We think of color mainly as an esthetic sense, but it is easy to understand why it would have had considerable survival value for the tree dweller. With a good sense of color the tree dweller could see brightly colored fruits, otherwise invisible against the leafy background, and was a better-nourished animal. Color vision also helped the tree dweller to pick out well-camouflaged predators who were difficult to see in the dappled shadows of the forest.

A better brain was also required for survival in the trees, in addition to the improvements in eye and hand. When the tree dwellers leaped from branch to branch, they needed a fast, accurate computer in their head to combine the factors of distance, wind speed, movements of branches, and balance of their body. The computation had to be done in a split second, with death or serious injury the penalty for any mistake. This natural computer – the brain – had to have a large memory capacity for storing the results of past experiences with aerial maneuvers; it also needed complex mental circuits to perform the necessary calculations; and it had to do its work rapidly.

The predecessor of the tree dweller, living on the forest floor, had possessed a brain without these qualities. But now, once again, the law of the survival of the fittest took effect. By virtue of the small variations from one individual to another that occur in every population, some animals possessed a slightly better brain than others; these animals were

FIGURE 14.8 Typical modern monkey from Asia.

© Joanna Van Gruisen/ardea.com

more likely to survive, passing on their superior brainpower to their offspring; those less well endowed with the necessary mental traits tended to perish at an early age, and their genes disappeared from the population.

Thus, under nature's pruning action, the brain of the tree dweller improved in quality and size; at the same time, the animal's dexterity and keenness of vision continued to develop. After 30 million years of improvements of body and brain, the descendants of the tree dweller had become another kind of creature. The new creature was the first monkey (Figure 14.8).

14.5 The evolution of monkeys and apes

The remains of the first true monkeys appear in the fossil record in rocks about 35 million years old. They were small squirrel-sized to cat-sized animals that apparently moved through the trees both by jumping and by running along the branches on all fours. The snout and nasal cavity of these animals were reduced in size, and the eyes faced directly forward. These ancestral monkeys had larger brains than the earlier tree dwellers. They were more intelligent than any other animal in their time, and became widespread and successful in the tropical forests of Africa and South America.

Not long after the first monkeylike creatures appeared in Africa, a few of their number apparently developed a novel pattern of behavior and associated anatomical changes. The unusual behavior involved the way in which these animals traveled in the forest. Some individuals discovered the trick of hanging by their arms and swinging from branch to branch, and even from tree to tree (Figure 14.9).

Hurtling through the forest canopy, the innovative primate could travel more rapidly than his cousins who ran along the branches. And the new method of travel had another advantage; tree dwellers accustomed to swinging by their arms could also distribute their weight among several branches, hanging from one overhead branch while standing on two others below. In this way, they could reach out with a free hand to gather ripe fruits inaccessible to other animals who could not accomplish this strategic trick.

Before that time, the danger of a branch breaking under the animal's weight had placed a severe constraint on the size of the tree dwellers; that is why monkeys were small when they first evolved, and most are still rather small today. (The largest monkeys are those, like baboons, that spend most of their time on the ground.) Now, some monkeys, swinging through the forest, could evolve into larger animals.

The fossil record reveals the result. Starting with an ancestral monkeylike creature that lived about 35 million years ago, the remains of the unusual group of tree dwellers show a progression in size during the next few million years. At the end of that time, the small monkey-like animals had been transformed into heavier-bodied creatures, weighing as much as 200 pounds (90 kilograms), with long arms and powerful shoulders, developed by evolution for ease and speed in swinging from one branch to another (Figure 14.10). But an animal with these traits of body and behavior was no longer a monkey; it was now an ape.

The apes underwent a burst of evolution about 20 million years ago in the widespread forests of the Miocene Epoch.

FIGURE 14.9 A chimpanzee eating figs while hanging from a branch and supporting its weight on two other branches.

James Moore, Anthrophoto

In another 10 million years, forerunners of all the modern branches of the family of the apes (and some that are now extinct) would be present: the agile gibbon, gymnast of the forest, a true ape of the trees, who has refined the arm-swinging mode of travel of the ancestral apes to an extra-ordinary degree; the orangutan, larger and heavier than the gibbon, but still primarily a tree dweller, who occasionally descends to the forest floor; the chimpanzee, bright, curious, spending much time on the ground, but at home in the trees

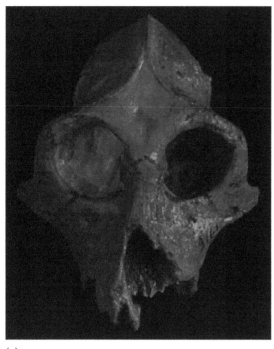

(a)

(b)

FIGURE 14.10 Skull and reconstruction of *Aegyptopithecus*, one of the earliest fossil apes. Fossils are known from Egyptian deposits about 35 million years old.

   Life Through Time by H. L. Levin. © 1975 William C. Brown Co.

and an agile arm-swinger; and the gorilla, a peaceful vegetarian of fearsome aspect, weighing as much as 600 pounds, who rarely goes up into the trees; and eventually, a new branch of creatures, more intelligent than any ape, but not yet human.

14.6 Summary

By the end of this chapter you should be familiar with the following concepts and topics:

Evolutionary radiation of the mammals
Success of the mammals
The evolution of mammal intelligence
The tree dwellers
The evolution of monkeys and apes

Questions

14.1. When did the first mammals appear in the fossil record, and what were their ancestors?

14.2. Why did the mammals undergo a major adaptive radiation beginning about 65 million years ago?

14.3. Why did increased intelligence develop in the mammalian branch of evolution?

14.4. What characteristics of the tree-dwelling mammals made them well adapted for life in the trees?

CHAPTER 15

The evolution of higher intelligence

Growth of the brain

Modern humans are the only creatures on Earth who have the intelligence and curiosity to wonder about their origins and their place in the Cosmos. To understand how we have become beings capable of such complex thought, we must look back in time to 10 million years ago, when a group of apes split off from the main branch of African forest apes and began to spend more time on the ground. For several million years thereafter, the record of human origins is rather sparse. But when the trail reappears in rocks about four and a half million years old, the apes had changed from a four-legged to a two-legged posture. What circumstances created this family of two-legged apes? What factors led to the subsequent explosive growth of the brain in the line of evolution leading to modern humans?

In this chapter we trace the family tree of human evolution from the early apes to modern man, beginning at the important point when apes started to walk on two legs instead of four. We look at the environmental changes that caused this bipedalism to occur, and what other advances became possible once the hands were freed from walking. We will see fossil evidence that shows how the ape body structure gradually changed through time into the upright stance of modern humans. Bipedalism allowed many other evolutionary developments, from the use of tools to the development of speech. We will consider these changes in relation to the dramatic increases in the size of the brain, and the advantages that they conferred. We then look at the eventual population explosion and wave of migration of fully modern humans, *Homo sapiens*, out of Africa and into Asia and Europe.

15.1 Evolution of bipedalism

An abundance of animal life flourished in the tropical African forests, including a variety of apes. About 10 million years ago, the climate in East Africa began to turn drier. It was also a time of global cooling. Evidence from study of the similarities and differences between DNA of modern humans and apes suggests that the split between the line leading to humans and the line leading to modern gorillas and chimpanzees occurred around that time.

Over time, because of the decrease in rainfall, dense forests gave way to more open woodlands, creating a rich mosaic of environments. Trees provided a safe haven, but some apes that spent more time on the ground would have been forced to travel from one heavily wooded area to another in search of food. A more upright posture would have freed the hands to carry food and ultimately for the use of simple tools. Yet, an ape on two legs is still an ape; its body is an ape's body. The awkwardness of the modern gorilla's posture is evident when it tries to stand on two legs. Its bones and joints are not built for an upright stance (Figure 15.1).

Every bone in the lower body of the ape had to change before it could stand with ease on two legs. The pelvis, thigh,

Previous page: Reconstruction of *Australopithecus africanus* in East Africa about 3 million years ago.
John Sibbick

FIGURE 15.1 A forest ape (the modern gorilla) attempting to stand upright.

leg, and foot acquired very different shapes (Figure 15.2). The shape of the foot is a clear example. In the ape, the foot looks something like a hand, and its big toe resembles a thumb; it is relatively short and angled out to the side.

By about 4 million years ago, the line leading to humans was clearly represented by *Australopithecus*.[1] The fossil remains of our early ancestor show that its big toe was large and pointed forward. It acted as a lever to provide a powerful forward thrust for the more humanlike stride of the two-legged animal (Figure 15.3).

Remarkable footprints of *Australopithecus* preserved in a layer of hardened volcanic ash about 3.6 million years old that were discovered by the late anthropologist Mary Leakey support this view (Figure 15.4). They are similar to footprints of modern humans.

An even more dramatic sign of the transformation from a four-legged stance to a two-legged stance can be seen in the skulls of *Australopithecus*. Every animal with a backbone has a conspicuous opening in its skull, called the *foramen magnum*, through which the nerves of the spinal cord pass on their way to the brain.

In humans, whose backbone is vertical, the spinal cord exits the head from beneath, through an opening in the bottom of the skull. In a four-footed creature like the dog or cat, whose backbone is horizontal, the spinal cord exits the skull through an opening at the rear. The fossil record shows that in the skull of the forest ape, the opening for the spinal cord was also toward the rear; but in the skull of *Australopithecus* it had rotated to a position almost directly underneath, and close to the location of this opening in the human skull. The position of that opening is evidence of a more erect posture (Figure 15.2).

Fossil evidence now shows that the earliest sites where bipeds are found were more wooded. This suggests that biped-alism was in place when the earliest hominids left the forest to explore more open areas. Bipedalism conferred significant

---

[1] Australo- means "southern"; -pithecus means "ape". As Australia is the southern continent, *Australopithecus* is the southern ape. It was given this name by Dr. Raymond Dart, because he found the first *Australopithecus* fossil in South Africa. Recently, the remains of a some-what earlier and more chimplike creature, named *Ardipithecus* (or *Australopithecus*) *ramidus*, have been discovered in sediments about 4.4 million years old.

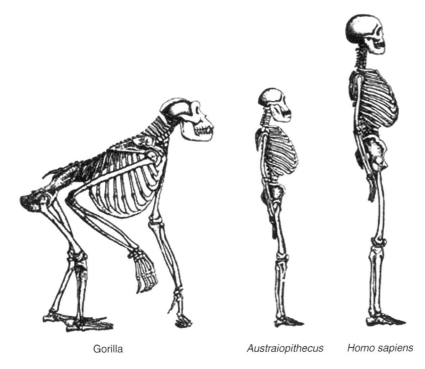

Gorilla   Austraiopithecus   Homo sapiens

FIGURE 15.2 Erect stance. The telltale sign of an erect stance is the point the spinal cord exits the skull, the *foramen magnum*. In the skeleton of the gorilla (left), the opening is toward the rear of the skull, indicating a four-legged posture. In *Australopithecus* (center), the opening is nearly underneath the skull, indicating a more upright posture. In humans (right), the opening is directly underneath.

The legs of *Australopithecus* are straighter and closer together than those of the gorilla; and the pelvis shows radical changes. In the ancestral forest ape and modern apes, the arms are relatively long in proportion to other parts of the body. In *Australopithecus*, the relative length of the arms is still somewhat greater than in *Homo sapiens*, suggesting that *Australopithecus* still retained some apelike features in its skeleton.

advantages in the open – it freed the hands to carry food and other items long distances.

Consider the application of Darwin's ideas to the forest apes who began to explore more open environments millions of years ago. These apes had a bodily structure that enabled them to stand erect more easily than others; a few, for example, possessed a thighbone and leg slightly longer than the average, giving them a larger stride.

These traits of the anatomy were favored or "good" characteristics under the pressures of life in the relatively open woodlands. In each generation, the apes with an erect posture left behind more progeny than their less favored neighbors; in the course of many generations, their progeny became very numerous; in time the entire population walked more erect.

The process may have been imperceptible from one generation to the next; but eventually the succession of many small changes made a new animal out of the old. During the long interval of many millions of years, these forest apes were transformed from quadrupeds to bipeds.

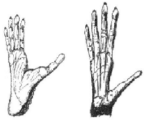

Gibbon or Tree ape

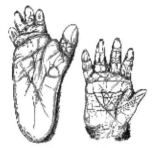

Gorilla or ground ape

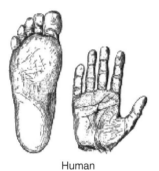

Human

FIGURE 15.3 Evolution of hands and feet. In the modern gibbon, which resembles the early tree apes, there is hardly any difference between hands and feet. The gorilla spends a great deal of time on the ground, and has a foot and hand that are quite different from each other. The gorilla's big toe is angled out to the side. The gorilla's hand is capable of neither a power grip nor a precision grip. In humans – and in our ancestors as far back as about 4 million years ago – all five toes point forward. These feet have apparently lost their grasping power. The hand has well-developed fingers and a large thumb.

15.2 The toolmakers

By 3 million years ago, the two-legged animals walked with a humanlike stride; their forelimbs were free for throwing stones and carrying food; they had a degree of manual dexterity; and they were accustomed to using their hands (Figure 15.5). Yet these were puny creatures, slight in build, 4 feet (1.2 meters) tall at most, weighing perhaps 70 pounds (30 kilograms), and no match for their competitors in physical prowess. Other animals had natural weapons – great strength, speed, size, claws that ripped, or slashing, stabbing fangs. *Australopithecus* had none of these. But this weakness was balanced by a well-developed brain and possibly the use of simple tools.

The size of the brain of *Australopithecus* was not impressive; at roughly 500 cubic centimeters, it was about as big as a fist and not much larger than the brain of the chimpanzee. But in proportion to body weight, the brain of *Australopithecus* was considerably larger than the brain of the ape. Every animal uses a part of its brain for receiving signals from the body and sending out messages in return; but *Australopithecus*, with a body considerably smaller than the ape's, required fewer brain cells for this purpose, and had more gray matter available for the storage of experiences from the past and the contemplation of actions in the future. Yet despite their seeming promise, these creatures became extinct about 1 million years ago.

Roughly 2.3 to 2.4 million years ago, well before *Australopithecus* vanished, another two-legged, intelligent animal appeared in Africa. The new animal was probably descended from *Australopithecus*, with similar bodily traits; but its brain was considerably larger (Figure 15.6). *Australopithecus* and its large-brained cousin lived on the same continent for almost one and a half million years. Throughout that long interval, while the brain size of *Australopithecus* remained largely unchanged, the brain of the other animal continued to grow.

The fossil record has not revealed why one cousin became more intelligent than the other; we only know that by the

FIGURE 15.4 Footprints of *Australopithecus afarensis* preserved in volcanic ash about 3.6 million years old in Tanzania in East Africa. The footprints show an almost modern stride. John Reader/Photo Researchers Inc. (FA0251)

(a)   (b)

time *Australopithecus* disappeared, its relative had acquired a brain nearly twice as large as the brain of the *Australopithecus*. This intelligent creature was the first animal to merit the designation *Homo*, a member of our own genus (Figure 15.7).

Signs of the superior intelligence of early *Homo* appear in the fossil record. Mary Leakey and Louis Leakey unearthed evidence in East Africa indicating that as early as 2 million years ago, *Homo*, with a brain size of about 600 cubic centimeters, was the master of a new toolmaking technology (the Leakeys named this fossil *Homo habilis*). Stone was the material used in this industry. Recent discoveries suggest that the manufacture of stone tools goes as far back as 2.5 million years.

The earliest stone tools made by early *Homo* were very crude, just pebbles broken in two to form a sharp edge (Figure 15.8(a)). Later examples had nicely chipped edges produced by a dozen well-placed blows, and made good cutting tools for butchering and skinning game (Figure 15.8(b)). Some tools have clean edges and look hardly used; in others, the cutting edge is battered and blunted by the wear and tear of heavy use. Tools that look like cleavers and chisels are also abundant.

The variety of tools found by the Leakeys and others is impressive. Even more impressive was the discovery that the materials used by early *Homo* in the toolmaking industry were

(a)

(b)

FIGURE 15.5 The evolution of the *Australopithecines.* (a) An earlier form, named *Australopithecus afarensis,* ranged from about 4 to 3 million years ago. (b) A later form, *Australopithecus robustus,* lived from about 2 to 1 million years ago. *Australopithecus robustus* had a much stronger jaw musculature and large chewing teeth. The powerful jaw muscles suggest that these pre-humans were adapted to a diet of tough vegetation.

not available in their campsites; they were a particularly hard kind of rock that was carried there from other places, in some cases as far as 10 miles away. Apparently, *Homo* scoured the neighborhood for miles around in the search for these stones, and brought them home. The implication is that *Homo* had developed an industry. These creatures seem to have thought out their needs in advance, gathered materials, and fashioned their tools.

The ability to make tools was once considered one of the characteristics that distinguished humans and our ancestors from all other animals. In 1964, Jane Goodall shattered this belief when she observed that chimpanzees in the African forest frequently make simple tools for catching termites (Figure 15.9). The ape first looks for the right materials; carefully selects a twig with the correct size and shape; then works on it, stripping off the leaves. When the tool is ready, the ape inserts it into a hole in the termite nest. Pulled out, the twig is

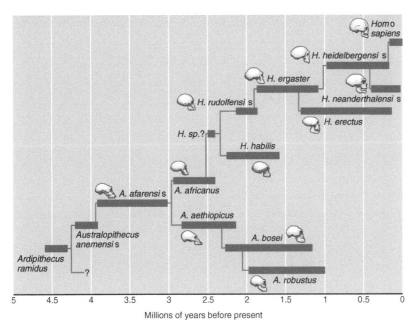

FIGURE 15.6 The family tree of human evolution. Several different species of *Australopithecus* are recognized prior to their extinction about 1 million years ago. The earliest members of the genus *Homo* date from around 2.5 million years ago. *Australopithecus* and *Homo* coexisted in Africa for about 1.5 million years.

Johanson and Edgar (1996) *From Lucy to Language*, 1996, Simon & Schuster

FIGURE 15.7 Early *Homo* had a larger brain (about 600 to 800 cubic centimeters) than *Australopithecus* (about 500 cubic centimeters), and produced stone tools.

(a)

(b)

FIGURE 15.8 Stone tools made by *Homo habilis*. The chipped stone (a) is a so-called pebble tool, made about 2 million years ago. The tool at right (b) is a cleaver, with a handle at the bottom. These tools, found in the Olduvai Gorge in East Africa, are among the earliest tools known.

covered with delectable insect morsels. Goodall also observed that this toolmaking technique is passed on from generation to generation among the chimpanzees. The young ones watch their elders and try to imitate them, clumsily at first and with greater skill later on.

If apes make tools, why are the tools of early *Homo* so remarkable? One reason is that the chipped rocks, which seem so primitive to us, are finely crafted instruments in comparison to the ape's termite stick. Another reason is that the ape takes its tools only from the materials at hand, and only for immediate use; it does not plan for the future. No chimpanzee has ever been observed to collect twigs on Monday, prepare them for termite fishing on Tuesday, and use them on Wednesday and Thursday. The contrast with the behavior of *Homo* is evident.

The toolmaking industry, judged in the context of the time, represented a high level of organization and planning. In intellect, this creature was superior to all other animals in its day; yet nature's work on its brain was far from complete. Early *Homo* was a nearly finished creature from the neck down, but its small skull held a brain half the size of the brain of modern humans.

15.2.1 Explosive growth of the brain

Early forms of *Homo* were followed by the appearance of *Homo erectus* about 1.5 million years ago. In *Homo erectus*, the brain size reached about 800 to 900 cubic centimeters. *Homo erectus* produced more sophisticated tools than early *Homo*, and may have used fire as early as 1.3 million years ago. They spread from Africa to Asia by about 1 million years ago.

Starting about a million years ago, the fossil record shows an accelerating growth of the human brain. It expanded at first at the rate of one cubic inch (a heaping tablespoonful, 15 cubic centimeters) of additional gray matter every hundred thousand years; then the growth rate doubled; it doubled again; and finally it doubled once more. Five hundred thousand years ago the rate of growth hit its peak. At that time the brain was expanding at a phenomenal rate of ten cubic inches (about 150 cubic centimeters) every hundred thousand years (Figure 15.10).

FIGURE 15.9 A chimpanzee with a termite stick. Chimpanzees commonly fashion and use simple tools. Recent studies suggest that gorillas and orangutans also make and use tools. Honolulu Zoo (www. honoluluzoo.org).

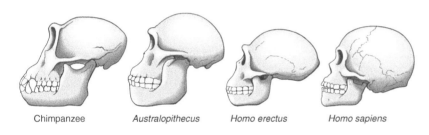

| Chimpanzee | Australopithecus | Homo erectus | Homo sapiens |

FIGURE 15.10 The forest apes of 10 million years ago had a skull resembling that of the chimpanzee (about 400 cubic centimeters in volume). The brain of *Australopithecus*, about 4 million years ago, was only moderately larger than the brain of the chimpanzee, but significantly larger than the chimpanzee's in proportion to body weight. The brain of *Homo erectus* (about 800 cubic centimeters) was considerably larger than the brain of *Australopithecus*. The earliest examples of even larger brained *Homo sapiens* (with a brain volume of more than 1200 cubic centimeters) appeared about 100 000 years ago.

15.2.2 Why did the brain grow rapidly?

What pressures generated the explosive growth of the human brain? A change of global climate that set in about 2 million years ago may supply part of the answer. At about that time, the world began its descent into a great Ice Age, the first to afflict the planet in hundreds of millions of years. The trend toward colder and drier weather set in slowly at first, but after a million years patches of ice began to form in the north. The ice patches thickened into glaciers as more snow fell, and then the glaciers merged into great sheets of ice, as much as 2 miles thick. When the ice sheets reached their maximum extent, they covered two-thirds of the North American continent and a large part of Europe. So much water was locked up on the land in the form of ice that the level of Earth's oceans dropped by 300 feet, and many regions, even in the tropics, became cooler and drier.

Moreover, the Ice Age has been marked by numerous oscillations of global climate, with alternating glacial and interglacial periods, and shorter intervals of a thousand years or less with quite sudden climatic flips. Periods of cold global climate led to severe drought in the tropics. Thus, each time the climate flipped to a cold phase, humans in the tropics were put under intense selective pressure. The turn toward a harsher and more variable climate coincided with the period of most rapid expansion of the human brain, suggesting a connection between the pressures of survival in changing times and the improvements in the brain of an animal that was increasingly living by its wits.

The toolmaking industry of early humans also stimulated the growth of the brain. The possession of a good brain had been one of the factors that enabled *Homo* to make tools at the start. But the use of tools became, in turn, a driving force toward the evolution of an even better brain. The characteristics of good memory, foresight, and innovativeness that were needed for toolmaking varied in strength from one individual to another. Those who possessed them in the greatest degree were the practical heroes of their day; they were likely to survive and prosper, while the individuals who lacked them were more likely to succumb to the pressures of the environment. Again these circumstances pruned the human stock, expanding the centers of the brain in which past experiences were recorded, future actions were contemplated, and new ideas were conceived. As a result, from generation to generation the brain grew larger.

The evolution of speech may have been the most important factor of all. When early humans mastered the loom of language, their progress accelerated dramatically. Through the spoken word, a new invention in toolmaking, or a new social development could be communicated to everyone; in this way the innovativeness of the individual enhanced the survival prospects of its fellows, and the creative strength of one became the strength of all. More important, through language the ideas of one generation could be passed on to the next, so that each generation inherited not only the genes of its ancestors but also their collective wisdom, transmitted through the magic of speech.

A million years ago, when this magic was not yet perfected, and language was a cruder art, those bands of humans who possessed the new gift in the highest degree were strongly favored in the struggle for existence. But the fabric of speech is woven out of many threads. The physical attributes of a voice box, lips, and tongue were among the necessary traits; but a good brain was also essential, to frame an abstract thought or represent an object by a word.

Now Darwin's law began to work on the population of early humans. Steadily, the physical apparatus for speech improved. At the same time, the centers of the brain devoted to speech grew in size and complexity, and in the course of many generations the whole brain grew with them. Once more, as with the use of tools, reciprocal forces came into play in which speech stimulated better brains, and brains improved the art of speech, and the curve of brain growth spiraled upward.

Speech also meant that humans could interact more closely in social groups. Social interaction requires the ability to read another individual's motives from their facial expressions, intonation, and behavior. These circumstances also fostered the growth of the brain.

Which factor played the most important role in the evolution of human intelligence? Was it the pressure of changing climates? Or tools? Or language? No one can tell; all worked together, through Darwin's law of natural selection, to produce the dramatic increase in the size of the brain, and the improvements in technology and culture that have been recorded in the fossil record in the last million years (Figure 15.11).

The earliest humans that populated northern Europe from 100 000 to 35 000 years ago are considered a separate species, *Homo neanderthalensis* (or Neanderthal man). These early humans were stocky in build as a result of adaptations to the harsh Ice Age climates (Figure 15.12), but their brains were essentially the same size as modern humans. Genetic evidence suggests that *Homo sapiens* and *Homo neanderthalensis* did not interbreed. Apparently, they led separate existences for thousands of years.

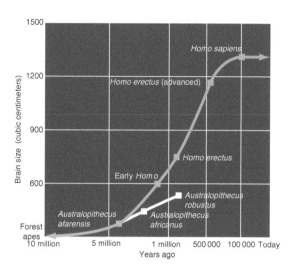

FIGURE 15.11 The fossil remains of our recent ancestors show that the brain began to grow rapidly about 1 million years ago, after a long period of moderate growth in the monkey and ape line of evolution. *Australopithecus* coexisted in Africa with early *Homo* until about 1 million years ago, when our smaller-brained cousins became extinct.

FIGURE 15.12 Neanderthal man (skull and wax replica). *Homo neanderthalensis* had a brain about the same size as modern humans.

© Philippe Plailly/Eurelios/Look at Sciences. Reconstruction by Atelier Daynes

Pre-modern *Homo sapiens* populated Africa prior to about 100 000 years ago.[2] Genetic evidence suggests that the human population was subsequently reduced to a few thousand individuals – a near extinction – possibly the result of a brief interval of severe climatic cooling and drought. *Homo*

[2] Fossils from Ethiopia discovered in 2003 indicate that early *Homo sapiens* emerged in Africa as early as about 160 000 years ago.

FIGURE 15.13 Cave paintings of horses, cattle, and other animals from a cave in southern Europe. These paintings were created by modern humans about 30 000 years ago.
   Norbert Aujoulat
© National centre of Prehistory, Montignac-Lascaux

neanderthalensis disappeared about 35 000 years ago, perhaps as a result of competition with *Homo sapiens*.

One hypothesis links the human population crash to a drastic climatic cooling triggered by the giant eruption of the volcano Toba in Indonesia about 73 000 years ago. Whatever the cause, this brief period of crisis was followed by a population explosion and a wave of migration of fully modern humans, *Homo sapiens*, out of Africa and into Asia and Europe.

These modern humans spread out of Africa some time around 50 000 years ago and brought with them the improved tools and culture that mark the final stages in the development of stone-age human societies. The remarkable cave paintings from southern Europe that date from 20 000 to 30 000 years ago show a rich artistic and cultural tradition that is clearly recognizable as the antecedent to our own (Figure 15.13).

During the past 40 000 years, humans spread to Australia, and to the New World by 12 000 to 15 000 years ago. The shift from hunter-gatherer societies to agriculture, the creation of new technologies, and the beginnings of modern human cultures all occurred within the brief 10 000-year span of our present warm interglacial period.

Today humans stand on Earth – intelligent and self-aware. We alone among all creatures on Earth have the curiosity to ask: How did I come into being? And, guided by our scientific knowledge, we come to the realization that we were created by all that came before us. In a 14-billion-year drama, the Universe has evolved from its formless beginnings into a complex tapestry inhabited by beings able to contemplate their long cosmic history.

15.3 Summary

By the end of this chapter you should be familiar with the following concepts and topics:

Human evolution: from early apes to modern man
Bipedalism
    Environmental changes that caused bipedalism to occur
    Other advances that became possible once the hands
       were freed
    Fossil evidence of the ape body structure modification
       to the upright stance
    Evolutionary developments allowed by bipedalism
    The use of tools
    The development of speech
The dramatic increases in the size of the brain and its
      advantages
The eventual population explosion
The wave of migration of fully modern humans from
      Africa to Asia and Europe.

Questions

15.1. List and explain the main steps in human evolution from the savanna apes to modern humans.

15.2. How much has the human brain grown during the last 4 million years, and what explanations have been suggested for the rapid growth?

15.3. When did *Homo sapiens* arise?

CHAPTER 16

Are we alone in the Universe?

Humankind's place in the cosmic community

Is life common in the Universe? Confirmation of the existence of life on Mars, Jupiter's moon Europa, or one of Saturn's moons would mean that life has independently evolved on more than one body in our Solar System and is not a highly improbable event. This would carry the implication that the Universe teems with life – much of it billions of years older than life on Earth. What is the chance that this life is intelligent? If the evolution of intelligence is a relatively common occurrence, where do we stand in relation to this cosmic community of intelligent beings? The history of life on Earth, combined with insights gained from astronomy, suggests some intriguing possibilities.

The chain of events that led to modern humans is hugely long and complex, and in this final chapter we question whether intelligent life may have evolved anywhere else in the Universe. We start by considering the existence of other habitable planets in the cosmos, and the possibility that life may have evolved there. We then move to the question of intelligent life. If other intelligent life does exist, where would we stand in relation to it, in terms of age and level of advancement? We will look at the probability that any other intelligent life forms would be anything like ourselves, and what this means for our potential to communicate with them. We will see which factors must be considered in trying to determine the number of other technological civilizations in our Galaxy. Finally, and perhaps most importantly, we will learn about the search strategies that are trying to detect the signals of other civilizations from Earth, and messages we have sent out into space in the hope of making contact.

16.1 Habitable planets

The imagination of the scientist has seized on bits and pieces of research accumulated in many different fields of science, and fashioned out of them a picture of the origin of life on Earth. No living form existed on our planet in its infancy and the atmosphere was filled with a noxious mixture of gases. In this environment, the basic building blocks of life were created. During the course of millions of years, the concentration of complex molecules increased; eventually a strand of nucleic acid appeared. Thus, the threshold was crossed from inorganic matter to a living organism.

According to this story, life can appear spontaneously on any earthlike planet that offers a congenial climate. How many such planets are there? The astronomical evidence suggests that there are billions of planets in our Galaxy, and billions more in other galaxies around us. In this great multitude of planets, some will be too close to their parent stars, and too hot; others will be too far away, and too cold; a few will be in the proper range to create the conditions needed for the evolution of life.

Previous page: Radiotelescopes used in the search for extraterrestrial intelligence.
© CSIRO, Australia

Let such favored planets be relatively few in number; let them be as rare as one in a million; no matter, the number of planets suitable for life would still be 100 000 in our Galaxy alone.

How many of these favored planets actually bear life? No one knows; all may be barren. But the biological discoveries described in this book suggest that this is not the case. Here is the evidence. All life depends on two-dozen molecular building blocks. Many of these building blocks of life are readily created in the laboratory under conditions that would prevail on any earthlike planet. The atoms that compose the building blocks of life are the same as the atoms that exist on every planet in the Universe; and there is reason to believe that the same laws of physics and chemistry apply in every corner of the Cosmos.

Why should the chain of chemical reactions that led to life on Earth not occur on planets circling other stars? Why should Earth – an undistinguished speck of planetary matter – be the only body in the Cosmos to bear life?

16.2 Are we alone?

The evolution of life out of non-life, even if scientifically explainable, could be an event of such vanishingly small probability as to be essentially a miracle. Now the significance of the discovery of primitive life on Mars or on some other body in the Solar System becomes clear. That discovery would indicate that the appearance of life on a planet *cannot* be a rare event. If it were, planets with life would be thinly scattered in the Universe and we would certainly not find two of these very rare objects – a planet with life – in the same Solar System. The discovery of life elsewhere in our Solar System would imply that many of the trillions of planets in the known Universe are inhabited. The Universe must be teeming with life.

16.2.1 What if life is common but intelligent life is rare?
Suppose life is common in the Universe but the evolution of a human level of intelligence is a fluke that has occurred only

rarely. According to the distinguished evolutionist Ernst Mayr, "millions of species get along fine without intelligence ... High intelligence is not at all favored by evolution." In other words, "smarter" is not better in a Darwinian sense, than "dumber."

Is "smarter" really no better than "dumber"? The fossil record reveals that some 99% of all forms of life have become extinct because the environment to which they have adapted has undergone a change, and bodily and behavioral traits, limited by the slow pace of natural selection, could not change rapidly in response. But intelligence is the best answer to this problem. Intelligence means the capability for a *flexible* response to changing conditions. And flexibility would seem to be uniquely valuable in coping with a changing environment.

The history of life supports the conclusion that intelligence is a valuable trait. The story of the evolution of intelligence begins 200 million years ago with the advent of the first known mammal – *Morganucadon*, the dawn mammal – a little rat-sized creature, but three to ten times brainier, pound for pound, than the ruling reptiles of the day. Over the course of 200 million years, the size of the brain relative to body size increased in the most intelligent mammals of each era – the ancestral monkeys, apes, and early humans – until, about 1 million years ago in this line of evolution, the brain exploded in size, roughly doubling in the next million years.

This is probably one of the fastest rates of growth of any organ in the history of life. Brains are heavy consumers of metabolic energy. Twenty percent of your blood supply is used in the brain. If big brains were not extremely valuable in the struggle for survival, natural selection would never have allowed these metabolically costly organs to evolve.

Mayr presented a second argument for the conclusion that human intelligence is unique in the Universe. He pointed out that the evolution of intelligence was the result of an almost unimaginably large number of branch points in the history of life. At each branch point, environmental pressures pushed evolution down one path when a different balance of pressures, on a different planet, could have pushed it down the other path. At the end of a long chain of evolutionary events on another planet, after a great number of different branches have been selected, the result is an entirely different creature. In the words of paleontologist Stephen Jay Gould, evolution depends on:

... a staggeringly improbable series of events ... utterly unpredictable and quite unrepeatable ... Wind back the tape of life ... Let it play again from an identical starting point, and the chances become vanishingly small that anything like human intelligence would grace the replay.

It is certainly correct that the probability of discovering another planet harboring *Homo sapiens* is essentially zero. But a close duplication of human evolution is not the issue. The question is not whether *Homo sapiens* or even a close facsimile has evolved elsewhere, but whether or not a human level of intelligence has been attained on other planets. Many distinct paths can lead to the evolution of intelligence, depending on the environmental history of a particular planet. That intelligence may be housed in bodies quite different from ours. Each of these evolutionary pathways to a human level of intelligence is of infinitesimally small probability, as the evolutionists point out. But the chance that at least one of those paths would have been followed is the sum of these many separate small probabilities. That sum can be nearly 100%.

16.3 Humankind's place in the cosmic community

If intelligence is common in the Cosmos, then we are only one among many intelligent forms of life. Where do we stand in this cosmic community of intelligent beings? Again we note the two vital numbers yielded by science. Without those two numbers, we know nothing about our probable place in the Cosmos. With them, we know a great deal. From astronomy, we have the number – 14 billion years – for the age of the Universe. From geology and the exploration of the Moon, we have a second number – 4.6 billion years – the age of Earth and the Solar System.

It follows from these numbers that we are recent arrivals in the Cosmos. Intelligent life in other solar systems, if it exists (for which we have as yet no evidence), is on the average billions of years older and more advanced than we are.

A billion years is a long time in biological evolution. One billion years ago, the fossil record tells us, the highest form of life on Earth was a wormlike creature. What stands in relation to humans as we stand in relation to the worms in the garden? The message in these scientific findings is clear. While humans stand at the summit of creation on this planet, in the cosmic order our position may be humble.

Will the extraterrestrials be humanoids – say, as humanlike as the creatures in Star Trek? Is a roughly human form the rule in the Cosmos or the exception? The history of life on Earth suggests an answer to this question. First, consider the events that led to humans on Earth. Members of our genus *Homo* appeared on Earth about 2 million years ago. They had evolved over the course of the previous million years out of *Australopithecus*, a creature that was half-human, half-ape with a humanlike body, but a brain that was one-third the size of the human brain. The ape-human had evolved in turn out of a chimpanzee-like forest ape some 6 million years before that.

So the fossil record tells us that in the branch of evolution leading to humankind, the rule is: A major change – and a major advance in intelligence and brain power – every few million years or so.

What will happen to us when some millions of years have passed and the new and more intelligent form has evolved out of *Homo sapiens*? According to the fossil record, 99% of all the species that existed on Earth have become extinct. *Extinction is the normal way of life.* We can conclude that before another few million years have passed, *Homo sapiens* will have vanished, to be replaced by a new and still more intelligent being.

16.3.1 Intelligent life in the Universe

A million years ago according to the fossil record the highest and most intelligent forms of life were subhuman. Today they are human. In another million years, judging by this record of evolution, we will be gone and our descendants will be superhuman.

Now we have the facts in hand to answer the question: How many of those beings out there are likely to be humanoid? According to our experience on Earth, humanlike creatures exist on an earthlike planet for about a million years. But a million years is a very small interval of time compared to the life span of the Universe. Think of the history of the Universe as a long line, stretching for 14 billion years from one end to the other. In that long line, just one narrow slice of time a million years wide is the time in which humanlike creatures are likely to be found on a planet. What is the chance of finding a star whose intelligent creatures are just in that humanlike stage of development? Picking a star in the Universe at random, the chance of its intelligent creatures being humanlike is approximately 1 million years / 10 billion years or 1/10 000.

In other words, only 1 in 10 000 of the planets with life will contain creatures that we could call humanoid. The other 9999 will be either too primitive to be able to talk to us or so advanced that we may not be able to talk to them.

Thus, it is exceedingly unlikely that any society on another planet came into existence at the same moment of time and developed at the same rate, so as to have arrived at precisely the same level of technology that we possess on Earth today. A difference of 100 years, which is the blink of an eye in the lifetime of a star or planet, has produced enormous changes in the scientific knowledge of our society. Some of these extra-terrestrial societies, living on planets born later than the Sun, must be primitive in comparison with us. Others, with an earlier start, must have surpassed our achievements a long time ago.

It is to this latter group – the more advanced societies – that we should direct our attention, for we must expect that they will have mastered the techniques of radio communication and harnessed the power required for transmitting signals over great distances, with a greater skill than we can hope to achieve in this century.

16.4 The search for extraterrestrial intelligence

While we have no idea of the number of inhabited planets in the observable Universe, we cannot, on the basis of facts available to us, exclude the possibility that such planets are common. Nor can we exclude the possibility that some of these inhabited planets harbor beings with intelligence comparable to or greater than human intelligence. We can be fairly sure, however, that if such beings exist, they will find it of interest to exchange radio messages with their neighbors; or they may be able to exchange information in the visible band of wavelengths using a punctuated code of laser pulses.

As we have seen, the majority of such beings or societies are apt to have a level of intelligence greater than ours. If they exist, they may have much of interest to tell us.

16.4.1 The Drake Equation

Radio astronomer Frank Drake of Cornell University proposed an equation that summarizes the information required to determine the number of technological civilizations that might exist in our Galaxy. If we represent the number of civilizations in the Galaxy as N, then the Drake Equation is written as:

$$N = R_* \cdot f_p \cdot n_e \cdot f_l \cdot f_i \cdot f_c \cdot L.$$

In this equation, R_* is the rate of star formation in the Galaxy, expressed in stars per year. This is the only parameter in the Drake Equation that is known with some confidence. Recent astronomical results indicate the current number may be as high as 10 to 20 stars per year. When the Universe was about one-third its present age, however, the star birthrate was substantially higher than it is now. This would mean that most of the sunlike stars in the Galaxy are billions of years older than the Sun, and that our civilization may be a relative latecomer. Not all stars, however, are equally well suited for the development of life. Very small stars, although long-lived, may be too dim keep their planets from freezing, and massive stars explode as supernovas after a few tens of millions of years.

The factor f_p represents the fraction of stars that have planets around them. Although most stars are binary or multiple systems, planets that form in single-star systems (like our own) are more likely to have stable orbits. It is widely thought that many single sunlike stars have planets. This is supported by the recent discoveries of planets around more than 100 nearby stars.

The factor n_e represents the average number of planets around a star that are capable of supporting life. In our Solar System, there are two possibilities: Earth and Mars. Two satellites of the giant planets, Europa and Titan, are also promising. But is our Solar System typical? All the other solar systems that have been found thus far have large, gas planets circling very close to their parent stars. It is possible that the formation of smaller, earthlike planets such as those in our own system is a rare event. On the other hand, computer simulations of planet formation from a gaseous nebula commonly produce several planets in earthlike orbits.

The fraction of those planets where life actually occurs is represented in the Drake Equation by f_l. It now appears that life on Earth arose rapidly once the planet had cooled enough to allow liquid water to accumulate on its surface. This is circumstantial evidence that planets with liquid water will eventually develop life.

The fraction of life-bearing planets on which intelligence arises is represented by f_i. On some planets, life may develop, but never reach the stage of evolution of complex organisms. This may have happened on Mars. On Earth, it took 4.6 billion years to produce intelligent beings, but on other planets the timescale might be significantly longer or shorter.

Some scientists reason that intelligence, having adaptive value, will eventually evolve on any planet where life arises and that maintains a relatively stable climate for a sufficiently long time. Thus, some scientists take f_i to be one.

The parameter f_c represents the fraction of planets with intelligent life where the intelligent beings develop the ability and willingness to communicate across interstellar distances. There may be planets where intelligent life arises, but never develops a technological civilization, or civilizations that do not attempt extraterrestrial communication for a variety of reasons. Imagine a planet where intelligent life takes the form of dolphin-like creatures that live in the sea. They might be intelligent and communicative, but they would lack the manipulative skills to build large radio transmitters.

The most important parameter of the Drake Equation in relation to the search for extraterrestrial intelligence is L, the lifetime (in years) of a technological, communicative civilization. The longevity of a technological civilization depends on its ability to overcome significant long-term problems, such as nuclear war, planetary environmental degradation, or global natural disasters such as asteroid impacts. If L is typically short, then very few extraterrestrial signals will be detectable at any given time. But if technological civilizations persist for long periods of time – hundreds of thousands to millions of years – then there may be many worlds transmitting signals that we could detect.

Estimates of the various parameters in the Drake Equation vary widely. If we use $R_* = 10$ stars per year, and rather pessimistic values for the other parameters, including the assumption that civilizations are short-lived and survive for only 50 years: $f_p = 0.5$, $n_e = 2$, $f_l = 1$, $f_i = 0.01$, $f_c = 0.01$, and $L = 50$ years. Then:

$$N = 10 \times 0.5 \times 2 \times 1 \times 0.01 \times 0.01 \times 50 = 0.05.$$

This low number for N would mean that we are essentially alone in the Galaxy.

Alternatively, making somewhat more optimistic assumptions, especially that civilizations are relatively long-lived and survive for 100 000 years (a short period of geological time), we could use $R_* = 20$ stars per year, and $f_p = 0.1$, $n_e = 0.5$, $f_l = 1$, $f_i = 0.5$, $f_c = 0.1$, and $L = 100\,000$ years. In this case:

$$N = 20 \times 0.1 \times 0.5 \times 1 \times 0.5 \times 0.1 \times 100,000 = 5000.$$

If 5000 advanced communicative civilizations exist in the Galaxy at present, then we can estimate the average distance between those civilizations. In our portion of the Galaxy, the stars are spaced about 5 light-years apart. Thus, there is about one star for every 500 cubic light-years of space.

There are about 400 billion stars in the Milky Way Galaxy, so that 5000 stars with planets harboring advanced civilizations would mean about one in every 80 million stars has a planet with an intelligent civilization. We would require a volume of 24 billion cubic light-years to include the nearest star with an advanced civilization. The distance to the nearest such star would be more than 1800 light-years.

Two problems are immediately apparent. The great distances between advanced civilizations, even in this relatively optimistic formulation of the Drake Equation, mean that a two-way conversation with our nearest neighbor civilization would take about 3600 years. The other point is that finding that civilization among 80 million stars would be looking for a needle in a very large haystack.

16.4.2 The Fermi Paradox

If civilizations last for relatively long periods, then we are faced with a problem that has been called the *Fermi Paradox* after the Italian physicist Enrico Fermi. After considering the possibilities of extraterrestrial intelligence and interstellar travel, Fermi asked, "Where is everybody?" His reasoning was that any relatively long-lived technological civilization should have eventually reached a stage where they initiated interstellar space travel. Even at sub-light speeds, such a civilization would have been able to colonize the entire Galaxy in a few tens of millions of years, a relatively short time compared to the age of the Universe. Since our civilization must be among the youngest in the Cosmos, many civilizations would pre-date ours by millions or even billions of years. We should be surrounded by galactic colonists.

Possible solutions to the Fermi Paradox are: (1) we are alone; (2) intelligent civilizations do not engage in interstellar travel and colonization, either because of technological difficulties, lack of initiative, or self-destructive tendencies; or (3) the Galaxy has been colonized, but the advanced civilizations are avoiding contact with us, or are too advanced to be detected.

16.4.3 Search strategies

The only way to know for sure if other technological civilizations exist is to detect their signals. With that thought in mind, beginning in the 1960s groups of astronomers and physicists set up projects designed to have some of the time on large, sensitive radio antennas, and Earth began to listen. So was born SETI – the Search for Extra Terrestrial Intelligence.

In 1960, in one of the earliest SETI studies, Frank Drake aimed an 85-foot (28-meter) antenna in the direction of two sunlike stars and listened in the microwave part of the radio spectrum. No extraterrestrial signals were detected. This first effort was followed by a number of projects in Russia and the United States. In the past 40 years, there have been more than 70 individual SETI projects.

Astronomers set up a listening post in 1995 to monitor 710 nearby stars with mass and other properties resembling the Sun. The privately funded program was named the Phoenix Project. In 2004, after nine years of listening, Frank Drake, one of the founders of the project, announced, "We found nothing." Plans call for expanding the project to listen to millions of stars in our neighborhood. The search continues.

16.4.4 Transmitting signs of intelligence on Earth

What about transmitting our own signals to the stars? In 1974, the most powerful broadcast ever beamed into space was made using the giant Arecibo Radio Telescope in Puerto Rico. The message consisted of a series of binary digits – zeroes and ones – arranged into a pictorial representation of our Solar System, the DNA molecule, a figure of a human, and the atoms of the elements critical for life. The target of the message was the globular star cluster M13, about 21 000 light-years from Earth. The problem is that the reply, if it ever comes, will take 42 000 years to reach Earth.

Since the 1960s, television stations scattered across Earth have been spraying their signals into space at a million-watt level. In the course of those 40 years, the expanding shell of television signals has swept past dozens of stars like the Sun. These television programs make Earth the brightest radio star at FM frequencies in our neighborhood of the Galaxy. Old programs, moving away from Earth at the speed of light, may have carried the message that intelligent life exists on this planet.

16.5 Summary

By the end of this chapter you should be familiar with the following concepts and topics:

Is there an intelligent life anywhere else in the Universe?
> The existence of other habitable planets
> Intelligent life
>> Does it exist anywhere else?
>> Where would we stand in relation to it in terms of age and advancement?
> Search strategies to detect signals of other civilizations
>> The Drake Equation
>> The Fermi Paradox
>> Messages sent out into space in the hope of making contact

Questions

16.1. What is the Drake Equation?

16.2. In the Drake Equation, f_i may be limited in part by asteroid impacts that would be catastrophic for technological civilizations. Civilizations are vulnerable to an "impact winter" from dust clouds produced by impact of asteroids larger than 1 kilometer in diameter. Such impacts are expected to occur on average about every 10^6 years.
How would these impacts affect the f_i term in the Drake Equation?

16.3. According to the accepted theory of planet formation, should planets be rare or abundant in the Cosmos, and why?

Epilog

Origins of Life in the Universe has followed the course of events in the evolution of the Cosmos from the Big Bang to the appearance of intelligent life. Questions of the origin of the Universe and its ultimate fate, the basic structure and origin of matter, the formation and distribution of earthlike planets, the origin and evolution of life and intelligence, and the occurrence of intelligent life on other planets in the Cosmos are among the most exciting lines of current scientific research. Astrophysicists are probing the first moments of the Universe's existence and speculating on the causes of the cosmic explosion; the Hubble Space Telescope and the best ground-based instruments are probing the limits of the observable Universe; new planets around nearby stars are discovered every month; the conditions for the origin of life are being investigated in laboratory experiments and in the exploration of the Solar System. Biologists and neuroscientists have even created an imposingly detailed picture of the workings of the human brain.

These areas of research pertinent to astrobiology provide information bearing on the possibility of life on other earthlike planets in the Cosmos. The probability of life and intelligence evolving out of non-living matter may be so small as to be a miracle – an event so unlikely that it has only happened once, or at most a few times, in the entire Universe. Or, it may be a likely event that has occurred on many of the countless billions of planets in the known Universe. The Universe may be densely populated with intelligent life, or we may be unique. In either case, the answer will have a major impact on our view of the significance of human existence in a cosmic setting.

Appendix A

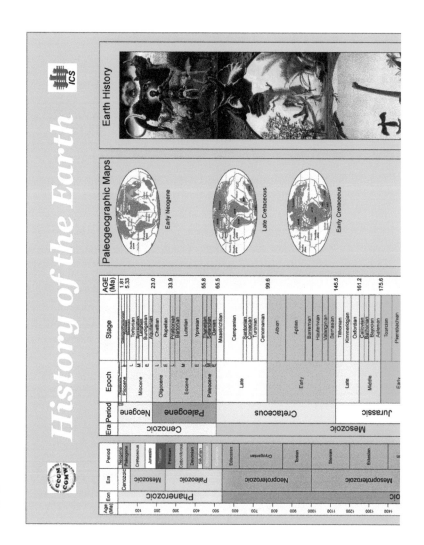

History of the Earth

Earth History

Paleogeographic Maps

Early Neogene

Late Cretaceous

Early Cretaceous

Era	Period	Epoch	Stage	AGE (Ma)
Cenozoic	Neogene	Pliocene	Gelasian / Piacenzian	1.81 / 5.33
		Miocene	Messinian / Tortonian / Serravallian / Langhian / Burdigalian / Aquitanian	23.0
	Paleogene	Oligocene	Chattian / Rupelian	33.9
		Eocene	Priabonian / Bartonian / Lutetian / Ypresian	55.8
		Paleocene	Thanetian / Selandian / Danian	65.5
Mesozoic	Cretaceous	Late	Maastrichtian / Campanian / Santonian / Coniacian / Turonian / Cenomanian	99.6
		Early	Albian / Aptian / Barremian / Hauterivian / Valanginian / Berriasian	145.5
	Jurassic	Late	Tithonian / Kimmeridgian / Oxfordian	161.2
		Middle	Callovian / Bathonian / Bajocian / Aalenian	175.6
		Early	Toarcian / Pliensbachian	

Eon	Era	Period	Age (Ma)
Phanerozoic	Cenozoic	Neogene / Paleogene	100 / 200
	Mesozoic	Cretaceous / Jurassic / Triassic	300
	Paleozoic	Permian / Carboniferous / Devonian / Silurian	400 / 500
Proterozoic	Neoproterozoic	Ediacaran / Cryogenian / Tonian	600 / 700 / 800 / 900
	Mesoproterozoic	Stenian / Ectasian	1000 / 1100 / 1200 / 1300 / 1400

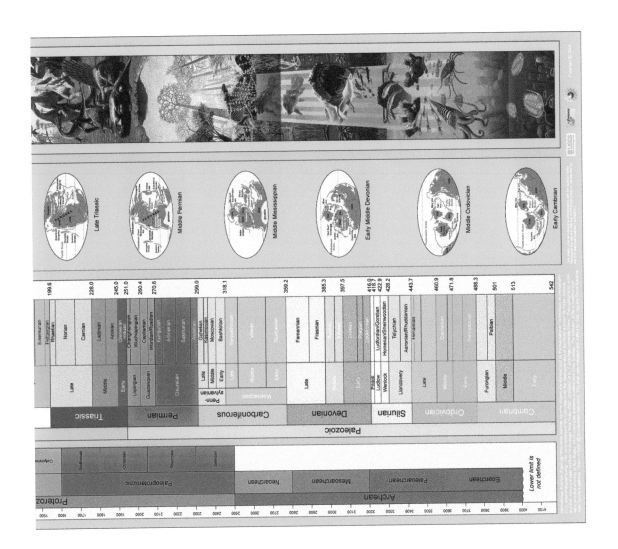

Appendix B

3 Elements and compounds

Key:

$\begin{array}{c} a \\ X \\ b \end{array}$

a = relative atomic mass
X = symbol
Name
b = atomic number

1
H
Hydrogen
1

	Group I	Group II												Group III	Group IV	Group V	Group VI	Group VII	Group 0
Period 1																			4 He Helium 2
Period 2	7 Li Lithium 3	9 Be Beryllium 4												11 B Boron 5	12 C Carbon 6	14 N Nitrogen 7	16 O Oxygen 8	19 F Fluorine 9	20 Ne Neon 10
Period 3	23 Na Sodium 11	24 Mg Magnesium 12												27 Al Aluminium 13	28 Si Silicon 14	31 P Phosphorus 15	32 S Sulphur 16	35.5 Cl Chlorine 17	40 Ar Argon 18
Period 4	39 K Potassium 19	40 Ca Calcium 20	45 Sc Scandium 21	48 Ti Titanium 22	51 V Vanadium 23	52 Cr Chromium 24	55 Mn Manganese 25	56 Fe Iron 26	59 Co Cobalt 27	59 Ni Nickel 28	64 Cu Copper 29	65 Zn Zinc 30	70 Ga Gallium 31	73 Ge Germanium 32	75 As Arsenic 33	79 Se Selenium 34	80 Br Bromine 35	84 Kr Krypton 36
Period 5	86 Rb Rubidium 37	88 Sr Strontium 38	89 Y Yttrium 39	91 Zr Zirconium 40	93 Nb Niobium 41	96 Mo Molybdenum 42	– Tc Technetium 43	101 Ru Ruthenium 44	103 Rh Rhodium 45	106 Pd Palladium 46	108 Ag Silver 47	112 Cd Cadmium 48	115 In Iridium 49	119 Sn Tin 50	122 Sb Antimony 51	128 Te Triturium 52	127 I Iodine 53	131 Xe Xenon 54
Period 6	133 Cs Caesium 55	137 Ba Barium 56	La to Lu 57	178 Hf Hafnium 72	181 Ta Tantalum 73	184 W Tungsten 74	186 Re Rhenium 75	190 Os Osmium 76	192 Ir Iridium 77	195 Pt Platinum 78	197 Au Gold 79	201 Hg Mercury 80	204 Tl Thallium 81	207 Pb Lead 82	209 Bi Bismuth 83	– Po Polonium 84	– At Astatine 85	– Rn Radon 86
Period 7	– Fr Francium 87	– Ra Radium 88	Ac to Lr															

139 La Lanthanum 57	140 Ce Cerium 58	141 Pr Praseodymium 59	144 Nd Neodymium 60	– Pm Promethium 61	150 Sm Samarium 62	152 Eu Europium 63	157 Gd Gadolinium 64	159 Tb Terbium 65	163 Dy Dysprosium 66	165 Ho Holmium 67	167 Er Erbium 68	169 Tm Thulium 69	173 Yb Ytterbium 70	175 Lu Lutetium 71
– Ac Actinium 89	– Th Thorium 90	– Pa Protactinium 91	– U Uranium 92	– Np Neptunium 93	– Pu Plutonium 94	– Am Americium 95	– Cm Curium 96	– Bk Berkelium 97	– Cf Californium 98	– Es Einsteinium 99	– Fm Fermium 100	– Md Mendelevium 101	– No Nobelium 102	– Lr Lawrencium 103

Elements in Groups I to 0 are sometimes known as the **main-group elements**

- The **reactive metals**: Group I – the alkali metals; Group II – the alkaline earth metals
- The **transition elements**: hard, strong and dense metals
- The **'poor' metals**
- The **metalloids**: includes semiconductors, e.g. silicon and germanium
- The **non-metals**: includes Group VII – the halogens
- The **noble gases**: very unreactive

Index

Page numbers in *italic* denote figures. Page numbers in **bold** denote tables.